Science,
Technology
and the
Human Prospect

Pergamon Policy Studies on Science and Technology

Gordon, Gerjuoy & Anderson *Life-Extending Technologies*
Hill & Utterback *Technological Innovation for a Dynamic Economy*

Related Titles

Bhalla *Towards Global Action for Appropriate Technology*
Bunge *Foundations & Philosophy of Science and Technology*
Constans *Marine Sources of Energy*
De Volpi *Proliferation, Plutonium and Policy*
Encel & Ronayne *Science, Technology and Public Policy: An International Perspective*
Garvey *Communication: The Essence of Science*
Miller, Suchner & Voelker *Citizenship in an Age of Science*
Williams & Deese *Nuclear Nonproliferation: The Spent Fuel Problem*

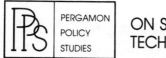 ON SCIENCE AND TECHNOLOGY

Science, Technology and the Human Prospect

Proceedings of the
Edison Centennial Symposium

Edited by
Chauncey Starr
Philip C. Ritterbush

Pergamon Press
New York □ Oxford □ Toronto □ Sydney □ Frankfurt □ Paris

Pergamon Press Offices:

U.S.A	Pergamon Press Inc., Maxwell House, Fairview Park, Elmsford, New York 10523, U.S.A.
U.K.	Pergamon Press Ltd., Headington Hill Hall, Oxford OX3 0BW, England
CANADA	Pergamon of Canada Ltd., 150 Consumers Road, Willowdale, Ontario M2J 1P9, Canada
AUSTRALIA	Pergamon Press (Aust) Pty. Ltd., P.O. Box 544, Potts Point, NSW 2011, Australia
FRANCE	Pergamon Press SARL, 24 rue des Ecoles, 75240 Paris, Cedex 05, France
FEDERAL REPUBLIC OF GERMANY	Pergamon Press GmbH, 6242 Kronberg/Taunus, Pferdstrasse 1, Federal Republic of Germany

Copyright © 1980 Pergamon Press Inc.

Back cover photograph: Thomas Alva Edison, as sculpted by Onorio Ruotolo, the only artist for whom the inventor ever posed in his Menlo Park laboratory. Photograph by permission of Professor Lucio Ruotolo, of Stanford University.

Library of Congress Cataloging in Publication Data

Edison Centennial Symposium, San Francisco, 1979.
   Science, technology, and the human prospect.

  (Pergamon policy studies)
  Bibliography: p.
  Includes index.
  1. Science—Social aspects—Congresses.
2. Technology—Social aspects—Congresses.
3. Inventions—Social aspects—Congresses.
4. Electric lighting—Social aspects—Congresses.
I. Starr, Chauncey. II. Ritterbush, Philip C.
III. Title.
Q175.4.E34   1979   301.24'3   79-18181
ISBN 0-08-024650-8
ISBN 0-08-024652-4 leather bdg.
ISBN 0-08-025595-7 pbk.

All Rights reserved. No part of this publication may be reproduced, stored in a retrieval system or transmitted in any form or by any means: electronic, electrostatic, magnetic tape, mechanical, photocopying, recording or otherwise, without permission in writing from the publishers.

Printed in the United States of America

Contents

Foreword—Observing the Centennial of Light—*Robert I. Smith*   vii

Introduction—The Growth of Limits—*Chauncey Starr*   ix

Part I: Judging the Costs and Benefits of Technology

1. Phases of Creativity in Science and Technology   *Gunnar Hambraeus*   3
2. Science, Technology, and Economic Growth   *Edwin Mansfield*   14
3. Science and Technology in Global Development   *Sumitro Djojohadikusumo*   26
4. Energy and Civilization   *George Basalla*   39
5. The Human Factor   *Eric Hoffer*   53

Part II: Adapting the Institutional Frame of Technology

6. Two Kinds of Light from Science   *Philip Morrison*   61
7. Technology and Socioeconomic Innovation   *Simon Ramo*   66
8. Public Reactions to Science and Technology: the Wizard Faces Social Judgment   *Jean-Jacques Salomon*   77
9. Industry and Energy: Moral Dimensions of the Tasks   *Alasdair MacIntyre*   94
10. Science, Technology, and Social Achievement   *Philip Handler*   109

Part III: Human Needs and the Future of Invention

11. Energy   *Wolf Häfele*   129
12. Technological History and Technical Problems   *Thomas P. Hughes*   141
13. Human Population and Ecology   *F. Kenneth Hare*   157
14. Medicine and Public Health   *Merril Eisenbud*   169
15. Urban Development   *John P. Eberhard*   177
16. Food and Agriculture   *René Dumont*   185

17. Democracy and Technology *Arthur Kantrowitz* (Convener)   199
18. Communications *Michael Tyler*   212

Index   221

About the Contributors   225

Foreword

OBSERVING THE CENTENNIAL OF LIGHT

This symposium on science, technology, and the human prospect has been a major event in the Centennial of Light, an international program observing the one-hundredth anniversary of the invention of the electric lighting system by Thomas Edison. Let us do more than honor the Edison spirit. The symposium convened in San Francisco from April 1-4, 1979, has the potential to bring new and creative ideas to play in a broad public debate on important questions about the future of industrial society.

Edison's work not only spawned the electrification of this country and the world but also inaugurated a century of technical and industrial development that has drastically altered our way of life and the way we look at life. This burst of technological development has brought comfort, wealth, and health to many. It has also left many untouched, out of the mainstream of material benefit. Will extending the applications of science yield even greater and more widespread human benefits? Or must we set a new course emphasizing nonmaterial values? Is it possible to pursue both objectives?

In the discourse that follows we will revisit the golden age of technology that poses such important choices for us. Our aim is not to promote technology for its own sake. We will find good and bad in our material development and its relation to the values of life. We approach the subject in a spirit of open inquiry and want as many people as possible to join in the dialogue. Decisions made in the next few years on directions for science and technology will influence our lives for decades to come. Accordingly we sought in the program of the symposium to canvass modern thought for wisdom to guide choices in the best interest of humanity.

I am pleased that Chauncey Starr, founding president and now vice-chairman of the Electric Power Research Institute (EPRI) introduced the proceedings. Throughout a distinguished career he has shown concern and sensitivity for the issues to be examined here. He has also formulated the program of the symposium, with the able assistance of Robert Loftness and a program advisory committee, working in close association with the institute's communication division, headed by Robert A. Sandberg, which has overseen the preparation for all symposium participants of a distinguished commemorative issue of the *EPRI*

Journal, "Creating the Electric Age: Roots of Industrial R & D," under the guest editorship of Nilo Lindgren.

Ten essays were commissioned from scholars and men richly experienced in the management of technology. These presentations form the first two parts of the symposium proceedings volume. The participants in the symposium also joined in policy review forums chaired by students of the relations of technology to social needs. These "workshops" met throughout the symposium to consider statements about problems and opportunities prepared by their presiding officers and other invited contributors. Their aim was to stimulate policy-oriented thinking about the issues addressed by the symposium. They appear below as commentaries, necessarily somewhat condensed, constituting the third part of the proceedings. We hope they may enlarge the reader's consideration of the themes developed in the commissioned essays. The names of those chairing the workshops appear as the authors of these chapters, while those invited by them to contribute papers are identified in the text and their contributions are identified by subheadings in each chapter.

Robert I. Smith

Introduction

THE GROWTH OF LIMITS

The coincidence of the one-hundredth anniversary of the electric light bulb with that of Albert Einstein's birth prompts reflection on the ways inventors and scientists have molded our society. Given that its framework is shaped by technology, we can't help but be concerned about how future technology will affect us. I will state my theme as follows: science and technology have continually relieved the limitations imposed by nature on man's living circumstances and will continue to do so. I believe this can be demonstrated from history.

Science and technology have permitted enormous increases in the world's population by improving man's ability to increase the food production of arable land; to increase the amount of productive land; to accommodate to harsh climates; to provide transportation and communication for the world's goods, services, and ideas; to increase available resources and use them more effectively; and to live longer in better health. Thus science and technology historically have opened new frontiers for mankind, not only permitting but also stimulating the "growth of limits." This phrase was suggested to me for use as my title by the study "Humangrowth" by Harlan Cleaveland and Thomas W. Wilson for the Aspen Institute for Humanistic Studies in 1978. I do not accept the thesis that constraints on growth are in view. Today's perceptions of limits are inadequate to determine our long-range planning horizons. A long future of expanding expectations continues to be available to us if we regard technology as an unlimited resource of the human mind.

Do we want to accept limits on human expectations and societal planning or do we want to keep alive the idea of expanding expectations? If we choose expansion we must have more science and technology applied to a spectrum of uses and addressed to the undesirable side effects of growth. The recently popular limits to growth theme was based on the assumption that natural resources and ecology impose a near-term limit on our ability to continue to derive materials from the earth and to handle with safety the residuals of growing industrial processes. For proponents of this theme, the obvious course would be to limit the personal economic expectations of those now alive and to sharply limit the growth in the world's population. This is a seductive dogma, because if it were socially and politically feasible to accept such limitations and thus reduce the worldwide demand for goods and services, we would simultaneously reduce the stress on our natural environment, on our political and social institutions, and on our problem-solving and decision-making mechanisms. But is such a prescription acceptable from a social and political standpoint? Perhaps it is for the comparatively wealthy, some of whom idealistically yearn for a simpler life or implicitly presume a resulting stabilization of their present comfortable status. But what about the poor people of the world, both those within the industrialized nations and the vast numbers in countries that have not yet caught up with the modern world? For them, limits to growth would mean lives of hopelessness and despair.

The continuing debate on energy supply and demand epitomizes these issues. Those who press the view that man-made energy processes alter the natural ecology appear to forget that human society also has a complex energy ecology woven into its fabric. As with natural ecology, any change in this total system of relationships will affect all its components. Popular ignorance of this fact has led to many simplistic notions on the role of energy in the quality of life, ranging from the idyllic notion that the less energy used, the better the life, to the opposite extreme, the need for unlimited abundance. For neither extreme can we find sufficient evidence. A grain of truth does not make a whole doctrine.

When we examine social groups with very low energy supplements we find they are frequently on starvation's edge, ridden with malnutrition, endemic disease, and physical misery. There is no mass migration into such deprived societies. It has been estimated that one-fourth of the world now lives on this edge of survival. Given a free choice, such groups avidly seek more energy input. It is clear from history, as well as from the visible living patterns in the range of societies worldwide, that a significant level of inanimate energy to supplement human labor does, by almost any humanistic measure, improve the quality of life. The vehicle for such energy input is technology. And this is intuitively understood by the world's peoples.

Thus, if personal aspirations worldwide cannot be reduced, limiting economic growth might indeed reduce the strain on our natural environment, but it would also create major—perhaps catastrophic—strains on our international political and social institutions.

I challenge the factual validity of the assumption that resources are limited, which is at the root of the doctrine of limited expectations. I believe that the resources available for use into the distant future are *not* generally limited because history testifies that advances in technology expand the availability of resources. Technology does this by providing increased efficiency in the conversion of resources to human uses (that is, less is needed to produce more) and in the extraction of traditional resources from the biosphere, as well as by providing methods for the conversion of dormant substances into new resources. So far, we have extracted only a small fraction of the store in the earth's crust. History also tells us that the apparently limited resources available to mankind in any one period of time become just a small fraction of the resources available to later generations because of the intervening contribution of science and technology. A key example is the relation of world food production to the population it can support. The following appraisal given in the *Scientific American* for December, 1922 illustrates this point:

> One of our leading statisticians estimates that a century from now our population will amount to more than 225 million people. The prophecy is startling because it suggests possible hunger and even famine as our future. At present, with less than half of these numbers, our food production is only about equal to our domestic consumption. Unless we institute very revolutionary practices to enhance production, we may look for it to fall behind.

This catastrophe, so obvious on this writer's assumptions, did not occur. As a result of improved technology in the agricultural sector we have instead become a major exporter of food.

As another example, consider the availability of oil. If oil exploration and extraction had been restricted to techniques current around 1900 we would have been out of oil a generation ago. At the beginning of this century, available United States reserves were about one billion barrels, enough for a decade then but sufficient now only for about two months of use. What oil was then known was found by looking for signs of seepage at the surface and by drilling around salt springs. In that period we didn't have the sophisticated exploration, deep drilling, and extraction techniques available today. Technology has enabled us to discover oil repositories miles underground and to force that oil out, as well as to obtain it through artesian pressures. And we've been able to open up oil fields under the oceans. There is no possible way that anyone without high-technology engineering equipment could develop a deep sea oil well. This is truly a frontier opened by technology and a good example of how technology has kept increasing the available supply of all mineral resources. This illustrates how science and technology are instrumental in expanding the limits of man's resources. This expansion of limits has not been just with minerals, but with food, with transportation, with communication, with health, and with extending livable space in unfriendly, harsh environments. And so, my keynote theme—the growth of limits—is both a statement of purpose and a subject of inquiry.

Let us now focus on the objectives of this symposium celebrating the Edison centennial, to reassess the role of science and technology in our social development, what they have done for the world's nations, what they mean to the individual, and where they are likely to go in the future. Such an assessment is particularly timely for the electric industry. The major economic growth of the industrial nations has taken place in parallel with and supported by the development of new uses for electricity, so the electric industry recognizes its responsibility as a partner in this growth.

Those of us pursuing the applications of science and technology believe that all progress in these fields is good for humanity generally and the developing nations in particular. Yet, we do recognize the serious negative impact that some of the by-products of these technical developments have had on individuals, on specific social trends, and on the environment in which we live. We also recognize the possibility that such social costs might become so large that we must carefully examine what we are doing and perhaps better plan our activities.

There has been—during this past decade, in particular—a growing attack by a portion of the intellectual community on the social merit of the industrial sector of our society. The electric utility industry is one such target, as are the automobile industry and the chemical industry. The natural-food movement is a popular reaction to the agricultural industry, which uses chemical fertilizers and pesticides heavily.

A segment of the intellectual community has suggested that we should somehow restructure our society to remove the heavy hand of technology. This has been one of the popular tenets of the environmental movement. Are we the passive subjects of some uncontrollable technical system or can we actively control this system to improve our destiny? I believe we must understand all aspects of this fundamental question and assist our society in finding a balanced answer.

Whether or not one pushes technological growth depends partly on individual philosophic beliefs and partly on the perceived prospects for technology. It's like sending your children to school, making a sacrifice to pay for their education. There is no guarantee of the long-range outcome. It is a matter of faith that their education is a valuable investment. For more than a hundred years most societies had unquestioned faith that technological growth was worthwhile. Our cultural conceptions of the role of science and technology are very fundamental to the average man. They affect his immediate activities, his future, and that of his children and grandchildren.

In the last few decades, our faith in technology has been questioned. And these doubts have created public issues bearing not only on the applications of science and technology but also on the creation of new knowledge. To some, the disapproval of the intended end use of a particular science or technology means that research to develop fundamental knowledge should be stopped. But to do so would inhibit the potential for its good use as well as bad. The debate on genetic engineering is such a case. In contrast, as an example of the stimulation fostered by an approved objective, the continual search for military superiority has historically extended scientific and engineering frontiers. It is an interesting commentary on the priorities of nations that this search for military strength has focused resources on new technology as no other social goal has done. It is indeed regrettable that our peaceful objectives are not equally stimulating.

As this symposium scans the activities of science and technology it will address the ethical issues that have been raised by the critics. We will not only examine some of the measurable values and costs but also the humanistic intangibles. How have the developments of science and technology changed the perception of the individual as to his role in society and his aspirations for the future? The key to this question is whether a scientific and technological age is perceived as serving or as thwarting basic human needs and aspirations. A correlated issue is the appropriate assignment of priorities in national planning. We have slowed the expansion of basic science and technology in the United States, restricting the amount of resources available for them. Simultaneously, we have thrown more resources into the intangible values of our society, such as expanding our scenic forests and improving the physical environment. How should our priorities be balanced? If, as many of us believe, the energetic application of science and technology in the past has provided enough economic surplus for both tangible and intangible uses and could continue to do so in the future, does the development and use of science and technology demand a higher priority than it is now receiving?

It is not sufficient that these matters be phrased in abstract terms and discussed only among those in the scholarly professions. Decisions on these important social and ethical issues clearly relate to the traditional values and objectives of the average man and to the common concerns of society, thus ultimately affecting the well-being of all. They concern the ability of society to provide abundantly for all human needs, material as well as spiritual, and to maintain a dynamism that leads to the development of new and improved useful products (as well as luxuries) over time.

Typical of the questions we are really asking when we plan our energy future is whether it is desirable to develop an artificial heart and other technical aids to keep someone alive or to delay death. Does this raise the issue of our right to interfere with nature? Should we modify natural environments to suit our perceived needs—create coal mines and build hydraulic dams? These are ethical—and important—issues. But, there is no one "truth" that will resolve such questions; for each we must seek a socially acceptable compromise.

My own recommendation is to strive for the greatest number of available options society can balance effectively. If someone wants to live in the redwoods and even if needed, wants to do without an artificial heart, I believe that should be an available individual choice. But if someone wants to live in the middle of San Francisco and wants to use an artificial heart so he can live there for ten more years, that option should also be available.

We are all exposed to propaganda urging particular views intended to modify our behavior. Perhaps all social systems require such indoctrination for internal self-discipline. However, there is a distinction between persuasion in an effort to achieve behavior modification and the restriction of options by prohibiting choice. I believe that persuasion and coercion are separable approaches. For me this is a matter of social ethics. Should society attempt to maintain freedom of choice or should it restrict options? It sounds very academic but it isn't. Everyone's daily life is affected by such philosophies. Indeed, they underlie our most pervasive political debates. It is this spectrum of issues that our symposium explores.

These issues are common to all societies, those affording economic incentives to individuals and those described as centrally planned economies. Social and political institutional mechanisms differ substantially among nations. Such differences importantly affect the daily lives of the people involved and could well be the subject of a separate symposium. But the questions regarding the role of science and technology that we here consider are universal and so arise in all economies.

I conclude by reemphasizing my own belief that science and technology are powerful and unlimited resources for bettering man's condition. The undesirable by-products of their application are amenable to reduction by these same arts. Science and technology are the key to the "growth of limits" for resources and to the application of these resources for useful purposes, as well as to the breaking of constraints on human expectations and aspirations. I hold to strong faith that future generations will be competent to manage the world they will inherit—one that I believe will be better than the world my generation inherited. I believe this can and will happen, provided our social and cultural structures do not inhibit their intrinsically fruitful activities.

Chauncey Starr

I

Judging the Costs and Benefits of Technology

1
Phases of Creativity in Science and Technology
Gunnar Hambraeus

Most of us believe that progress in science and technology affords the basis for prosperity and advancement of the quality of life. Creativity is important to everyone's future for it is the ability to set new patterns of thought, to identify new problems and find better solutions for old ones, and to discern new structures in matter. As this faculty is pecular to certain individuals and societies it should lie within our power to foster it. We may be baffled at first by its human uniqueness. After all, no animal has ever shown it. If, sometime in the future, computers will share our ability to set new patterns they will function within the limits set by their circuitry, just as animals are incapable of transcending the limits set by genetic structure.

In engineering we know we use only a small part of the theoretical strength of our materials. So human beings make scant use of their full intellectual faculties. Is creativity, then, the release of inner powers to some heightened extent? How often creativity is overwhelmed by circumstance! Évariste Galois, the founder of group theory, was 20 years old when he died in a stupid duel. Niels Henrik Abel, already famous for his work on 5th power equations, died of consumption at 27. Would a creativity policy serve, in principle at least, to counter the negative circumstances ever ready to encroach on the fertility of the mind?

Too much affirmation may not be good for creativity. The interactions with the outer world in which mental powers are released should not be entirely benevolent. Obstacles such as competition, stress, harsh demands, even compulsions, can release mental energies that might otherwise have lain dormant. Let us not lapse into the cynicism that excuses the shortcomings of our educational system on grounds that "nothing can stop talent." This would press a limited truth too far. Creativity requires a fair chance at least, even if its way need not always be smooth. Societies finding menace in the unfamiliar may resist new thoughts, and then creativity seems truly forestalled. From the Heliaster in Athens that condemned Socrates in the year 399 B.C. through the Inquisition that silenced Galileo to the Cultural Revolution in the People's Republic of China, the persecution of intellectuals by public authorities has been able to restrain the scientific mind. There have been other periods characterized by exceptional intellectual vigor. Recollections of periods when creativity burst forth with particular force may reveal general phases or patterns.

GREECE AND HELLAS: THE THINKERS

The origins of Greek culture are lost in ancient myth and legends. When written history begins, the Greeks had already settled the shores of the Mediterranean like "frogs around a pond," as Plato put it. Trade was intense. Wars were frequent. Elegant diremes and triremes provided easy travel over great distances. Three migration waves produced a genetic blend and a rich joint heritage from Egypt, Phoenicia, and Persia. In Miletos, Smyrna, Ephesus, the island of Samos, and, above all, in Athens, intellectual curiosity combined with youthful experimentation to lay foundations of thought for civilization in the West.

Thales from Miletos was a great mathematician who also knew how to apply his knowledge. Anaximander, Anaximenes, and Anaxagoras came closer to the structure of the solar system than other astronomers until Kepler's day. Aristotle, pupil of Plato and the teacher of Alexander the Great, thought and lectured on all philosophy and science. He became the absolute authority for almost 2,000 years. Modern logic as well as experimental physics still have their roots in his teaching, just as political theory springs from the discourse he and Plato shared.

The science of Hellas flourished for five hundred years, less exuberantly after the Macedonian conquest of the known world than before it, but enough to influence all future learning. In Alexandria the first scientific institution, the Museion, had a library of half a million volumes. Euclid and Diaphantos brought mathematics to a peak. Ptolemy developed his world structure and Hero experimented in mechanics and built intriguing if useless machines. The giant of applied mathematics, Archimedes, protected his ruler in Syracuse against both a dishonest goldsmith and the Roman general Marcellus.

ANCIENT, NONINVENTIVE, CHINA

On the other side of the globe in China, the fifth, fourth, and third centuries before Christ were ages of great political confusion. Only in the year 221 B.C.

was China united by the Emperor Cheng whose Chin dynasty gave the country its name. These tumultuous centuries of the warring states brought an upsurge in philosophy, a classical age of progress in science, particularly astronomy. Unlike Alexander, the Chin emperor did not enlist philosophers at his side and science suffered an irreparable disaster along with the rest of Chinese culture in the vast book burning in 213 B.C.

We may contrast the creativity of the Greek and Chinese cultures. Prosperity was probably higher in Honan and Shan-si than in the Peloponnesus. Travel in China was more extensive, in carts and chariots, or by boats on the rivers. Racial intermixture resulted from the hungry barbarians ever breaking through the frontier walls on their way to absorption by the vastness of the interior. Architecture and art flourished, but science developed only slowly. The astronomer Chang Heng could foretell eclipses; he was the first to observe sunspots and to build a seismograph. But it was around 500 A.D. before π was known with exactness. An understanding of the infinitesimal, which Democritos had reached already in the age of Plato, never entered the Chinese scholars' minds.

We are used to hearing recitals of the great inventions of the Chinese: the compass by Huang-ti and paper by Tsai-Lun some two hundred years before Christ; and gunpowder and printing by loose type about ten centuries after the beginning of our era. But it would be incorrect to speak of the Chinese as an industrially inventive people. They were creative in art, developing forms which were their own and unsurpassed in sensitive perfection. Agriculture, water management, and sericulture were highly developed at an early stage. But the Chinese remained content with ancient economic ways and scorned labor-saving devices. Mechanisms were not used in mines or the production of goods. Even war, the universal progenitor of technology, failed to spark inventions. It took the Chinese more than 200 years to find out that the most effective use of gunpowder was not for fireworks but in firearms.

ROME: THE ENGINEERS

The god of war, Mars, was the stern teacher of the Romans. They were creative in an unheroic, matter-of-fact way and excelled in what we now call civil engineering. The building of roads, bridges, aqueducts, walls, sewers, and fortifications, the walling of rivers, the draining of swamps, the dredging of harbors, and the construction of stupendous public baths and arenas were carried out by armies of free laborers, legionnaires, and slaves. Anonymous inventors created concrete and designed duct systems for heating and pipelines for water distribution. Mass production of weapons and household goods was organized, and Rome was fed by grain from Africa brought on bulky wheat ships, 180 feet long, with 44 foot beams.

The proconsuls at the head of the Roman legions carried their technology to the very end of the known world. The villa of an official in Bath in England was as well heated and supplied with running water as if he had been living only a few miles outside of Rome. A winter camp in Persia was laid out according to the same pattern and with the same sanitation and food factories as its counterpart on the shores of the Rhine.

THE PLODDING PROGRESS OF THE DARK AGES

Creativity and innovation became even more synonymous during the Middle Ages. It is wrong, however, to think of those times as lacking in progress. The productivity of agriculture increased with the invention of deep plowing, the horse-collar harness, the horseshoe, and shift cultivation. Mechanical clocks, sometimes of an astounding complexity, were built. Big organs were installed in beautiful cathedrals erected with unsurpassed skill from natural stone. Mining and metallurgy flourished. Canals with locks were built, and wind and water power were harnessed to power mills and smithies.

The stirrup made possible the combat of armored knights. In time the knights were overcome, first by the longbow and later by firearms. At sea the long ships of the Vikings appeared, superb sailing machines lighter, faster, and more durable than anything before.

A great innovation of enduring importance appeared in the eleventh century in Salerno, the medical school that became the model for all universities. In spite of scholastic formalism and clerical domination, these institutions of higher learning became the breeding grounds for new intellectual ventures. By the year 1500 there were 77 universities in Europe, of which the most famous was in Paris.

THE RENAISSANCE: SCIENCE REDISCOVERED

Nicholas Copernicus found an abstract means of simplifying the complicated Ptolemaic model of the planetary system, which forced him to the conclusion that the sun and not the earth stands at its center. Tycho Brahe's observatory on the Danish island of Ven gave Kepler the basis for his laws that the planets moved around it in ellipses. All this work was of a mathematical nature that did not disturb the authorities. Then Galileo with his telescope crushed the medieval crystal spheres. Through his glass anyone could see for himself that the teachings of the church were not true. A thin book of one hundred pages, *Sidereus Nuncius*, in which he described his observations, became the first scientific bestseller. It was torn from the hands of the printer and sent by special courier to all the courts of Europe, reaching distant Peking in a few years. Galileo can be seen as the symbolic peak of the intellectual revolt of the Renaissance.

Historians have debated the driving forces behind the exuberant vitality of the sixteenth century. Europe had recovered from the Black Death. New methods in agriculture and the emergence of large-scale manufacturing in cities as in the north of Italy provided great wealth. It had become the fashion of the rulers to employ artists for the beautification of their palaces, engineers for military construction, and scientists to mint coins and foretell the future. Conquerors were sent out to spread the Gospel and to find gold. The news services built by bankers to promote financial operations and commerce also spread new scientific knowledge. And the universities became focal points for debate and critique.

Above all, philosophy and theoretical science were again brought into contact with reality and practical problems. The threads left dangling at the death of Archimedes were taken up by Leonardo da Vinci and Galileo. They were concerned with problems of pumping water out of mines, of calculating the trajectory of a cannon ball or of laying out fortifications. The new scientists bridged the gap between the artisan and the scholar.

EUROPEAN SCIENCE: THEORIES AND EXPERIMENTS

Europe from the Renaissance up to our time shows in many respects a similarity to Greece in the five centuries before Christ. Instead of city-states we see nations; the Aegean Sea is replaced by wider oceans; and great wealth is created and commerce develops. Wars are endemic, new philosophies fight religion, and political experiments oscillate between tyranny and rampant revolution. With experimental proofs, after all, authority had become subject to challenge.

The masters of science become too numerous to tell. Only giants stand out. A modest Isaac Newton declared that if he had succeeded in seeing farther than Descartes, it was because he stood on the shoulders of giants. Newton is perhaps the best example of the inexplicable human phenomenon, the scientific genius. No outstanding student in grammar school, he came to Cambridge in 1660. Six years later he had developed his infinitesimal calculus, the theories of gravitation and the motion of the planets, as well as his optics. It then took him 20 years to explain and describe his discoveries.

Leonard Euler, mathematician, served the great and enlightened rulers of the eighteenth century, the Empress Catherine and Frederick the Great. Johann Friedrich Carl Gauss astounded his contemporaries by his published discoveries; and in the little notebook in which he sketched problems for future study, found after his death, he seems to have foretold most of the mathematics of the hundred years following!

In physics, electricity was discovered. Following a number of eminent experimentalists, Michael Faraday brought forward the theory of the electromagnetic field. It was perfected by James Clerk Maxwell in equations that still stand as the foundation of all things electric. Modern chemistry originated with Robert Boyle, Antoine-Laurent Lavoisier, and Carl Wilhelm Scheele. In 1808 John Dalton published his atomic theory. But only in 1869 was complete order given to chemistry in the periodic system of the Russian D. I. Mendeleev.

In the life sciences, Charles Darwin's *Origin of Species* opened the deepest schism between religion and science since Galileo. A century before Darwin, Carl von Linné in distant Upsala had by his *Systema Naturae* ordered the fauna and flora of the world. Later in the nineteenth century a monk unknown to the leaders of science, Gregor Mendel, experimenting with garden peas, discovered the laws of heredity.

Physics was held in focus largely by the efforts of a remarkable series of scientists at the Cavendish Laboratory in Cambridge. For many years this illustrious school dominated the roll of Nobel laureates: John William Strutt, the third Baron Rayleigh, Rutherford, John Thomson, the two Braggs, Chadwick, and Cockcroft. They represent a brilliant experimental tradition of symbiosis between theory and simple experimentation which was so typical of the Cavendish. Fondly called "the string and sealing wax method," it was nevertheless unsurpassed in revealing the deep secrets of matter and in providing a base for unified theoretical physics. This beautiful experimental art is now largely lost. With huge accelerators, big telescopes, radio discs, and space probes we have entered the age of "big science." Experiments take years and millions of dollars to prepare. A single scientist may make an experiment no more often than every five years. It is through sequences such as this in history that we realize how creativity lies open to the influence of institutions.

THE GOLDEN AGE OF INVENTIONS

A new age of inventions started with the nineteenth century. The Jacquard loom was exhibited in Paris in 1801. Robert Fulton's first commercially successful steamboat dated from 1807; Michael Faraday's dynamo from 1831; and Samuel Morse's telegraph from 1835. Charles Goodyear accidentally vulcanized rubber on his hot stove in 1839. Although he never admitted it, chance intervened in Alfred Nobel's search for a safe absorbent for nitroglycerin. From a leaky can some of the tricky fluid was spilled in a packing material, kiselgur, and dynamite was invented in 1866.

Alexander Graham Bell filed his basic telephone patent a few hours before a competitor on the morning of February 14, 1876. Three years later the incandescent lamp was born, again in a hectic race between several contestants. Thomas Alva Edison was the personified incarnation of the new breed of inventors.

Behind this surge of innovation was a persuasive idea. As Alfred North Whitehead would write, "The greatest invention of the nineteenth century was the method of invention." The economic rewards of success under the United States patent system were great and well advertised. Successful innovations created fortunes, though perhaps not always for their progenitors.

There was during these decades a fury for fame and fortune that our jaded twentieth century can hardly imagine. This provided a mighty momentum to creativity and ingenuity. It also released resources for research and development. Once it was proven that risk capital invested in experimental facilities for gifted technologists and scientists could give a huge return, the next logical step was the industrial research laboratory. Menlo Park was the forerunner to several generations of research and development institutes, culminating in the research parks of General Electric, Imperial Chemical Industries, and Philips of Eindhoven in the Netherlands.

The great asset of the large research laboratory is the possibility of bringing one single problem under a concentrated attack. Edison's search for a suitable material for the hairpin wire in the incandescent lamp is a good example of this. Hundreds of different organic fibers were carbonized until splints from a special bamboo proved to be the solution.

These brute-force campaigns are particularly efficient once the leading discovery is made. After Dr. Alexander Fleming had observed the antibacterial effect of *Penicilium notatum* the finding of new antibiotics and methods for their production became an ideal task for the research and development units of the pharmaceutical industry the world over. The search for new polymers after the invention of nylon is another example. The same process can be observed today with semiconductors and microelectronics. Once the road to large-scale integration was marked, the realization of the megabyte chip became a matter of money and a race with time.

CREATIVITY IN SCIENCE

The appearance of a genius presupposes a unique combination of genes. By pure chance an individual may have the qualities of an outstanding philosopher, a born mathematician, or an illustrious physicist. Under certain circumstances these qualities will be nurtured and developed. Good or bad schooling does not seem to play a big role nor does material wealth over a certain threshold.

One crucial factor seems to be the encouragement at the right moment of the curiosity, the inquisitive spirit, and the urge to explore the unknown that characterize the young scientist or innovator. Another factor is the early exposure to new ideas and intellectual debate. The intellectual input does not need to be of a special type. New thoughts in philosophy can provide stimuli for art, mathematics, or literature. In many cases the origin can be traced to the mixing of cultures, to wars and conquests, to new transportation and commerce, although it is sometimes difficult to decide whether a political disturbance is the sign or the cause of intellectual ferment.

To blossom creativity also needs a climate of understanding and appreciation. Society as such, a segment of society, or even one single person, such as the enlightened rulers in Europe during the eighteenth century, may provide such stimulus. In ancient Greece, it was the freemen, in Renaissance Italy the prince, and in nineteenth-century Europe the international scientific society that gave encouragement and incentives to intellectual pursuits. Very rarely was this type of creativity brought forth by the allure of wealth. It was, and still is, the respect and recognition of achievement, the prizes and honors bestowed and, above all, the inner satisfaction of an important task well fulfilled that spur the true scientist.

ALL THE SCIENCE'S MEN

Science is not only brilliant new concepts, theories that break frontiers, or the combining of facts into astounding new structures. It is also the slow and wearisome collection and sorting of facts. It is the designing, trial, and discarding of one hypothesis after another. It is the tidying and following up on the work of the trail blazers. This is the work of innumerable men of lesser or even humble status, and in total it is of no less importance than that of the brilliant masters. These, the scientific cadres, to use a Chinese phrase, are of even greater interest to our discussion today. Through a wise science policy we may encourage gifted young people to study science and technology, to stay with a career in research and development, and to add bit by bit to our knowledge.

In most industrial countries the last few decades have seen a gradually diminishing enrollment in the science faculties of the universities. This has been attributed to a negative attitude toward science and technology that has become general in society and perhaps more pronounced in the mass media. There are, however, a number of more rational reasons. New, and in some respects less difficult, lines of higher study have been opened, like social science and economics. Universities now offer a wider choice than before. The teachers in high school and elementary school have less knowledge, understanding, and training in science, and little or no experience of or connection with industry. The naturally inquisitive child entering school meets there little encouragement to ask questions on science and engineering, and receives little information on the physical and technical aspects of his immediate surroundings. Extracurricular activities connected with natural science used to play an important role in arousing interest. Now intelligent toys and hobby computers may take the place of radio tinkering. The importance of television in stimulating curiosity and inspiring a search for knowledge cannot be overestimated.

We must also acknowledge that many interests nowadays compete for a young person's time: more social activities, sports, work to earn money, and above all television. At colleges and universities the students are all too seldom exposed to the inspired teaching of the highly learned. Through the ages such exposure has been perhaps the greatest stimulus to academic pursuits. It is regrettable that the personal master-pupil relation cannot be maintained in modern university mass education. Moreover, the scientific profession is a demanding one. Not every young person who seeks academic training will prove himself as a creative research apprentice. There has to be, or has to be created, a wider employment market for university graduates and doctors than research and development itself can provide in teaching, civil service, industry, and journalism. It would be highly beneficial to the recruitment of students into the science faculties if extramural careers could be recognized as just as important as the academic ones.

The manpower of science is the most important ingredient of scientific progress. Adequate material support and efficient services nonetheless play a great role. Documentation, instrumentation—which can now include very heavy equipment indeed—workshops, skilled technicians, and good laboratory facilities are crucial to the success of much research.

This brings up the subject of money. Research and development has become big business. Very large appropriations are forthcoming from national governments. In many countries the public is the dominant supporter not only of basic science but also of much applied and industrial research. The allocation and distribution of these funds are sometimes of great political concern and increasingly handled by a new corps of professionals, research and development administrators. With rapidly rising costs in experimental science and engineering research the notion that intellectual pursuits enjoy unlimited academic freedom is largely illusory.

Possibly the political and administrative steering of research is more extensive than we care to admit. There are fashions in research as well as in art. Big established fields tend to draw big money long after the law of diminishing returns has made every new piece of knowledge excessively expensive. It is more convenient and safe for a scientist to stay with his specialty than to break new ground. Everyone who has once sat on a research council knows how difficult it is to terminate even obviously worthless projects. A much higher, if necessarily enforced, flexibility in the allocation of research funds would probably greatly stimulate creativity in science.

APPLIED SCIENCE AND INDUSTRIAL INNOVATIONS

So far I have mainly dealt with scientific creativeness, the origin of new knowledge. Let us now turn to the innovative process: the invention, development, and commercial introduction of new products and services. Innovations can be the results of scientific research. More often, however, they stem from a quite different background. Innovations usually start with the perception of something needed, with a function to be filled or improved, as well as the demand of a market.

Still, it is the original mind that perceives new possibilities in the uses of knowledge. In many respects the inventor, the first link in the innovation chain, is like the scientist. He is dependent on the attitude of his environment, the encouragement of colleagues, and the free exchange of information and ideas. In two respects, however, the typical inventor differs from his colleague in fundamental science. He is strongly stimulated by the possibility of economic gains and he is in a curious way independent of traditional schooling. The inventor has little need to collect knowledge in general, but once he has perceived an idea he can go to no end of trouble to find the theories and facts needed for the realization of his vision. This is not peculiar to the lone inventor. It has in fact become the traditional way of industrial development. First the functional needs are defined, then the search for solutions is started. In this process pertinent knowledge is applied. Certainly, the very highest learning and skill is found in the industrial laboratories. It is not, however, permitted the academic freedom to find its own outlets but rather strongly guided and bent to the program of a particular company.

The most important prerequisite for industrial development seems to be the awareness of the necessity of constant progress. If there is a demand for development at all levels in a company—and this rule holds also for the public sector and for society as a whole—there will not only be less resistance to change but there will also be an inner evolution resulting in rising productivity. Thus in innovation as well as in science, progress depends on a favorable attitude toward intellectual efforts, industrial development, and economic growth. Whenever civilizations have set other values above these goals, science and technology as well as industrial development have suffered. This is the fundamental explanation of the slow growth or stagnations in Europe during the Dark Ages, in China and Japan up to the twentieth century, and in India until its independence.

I have repeatedly mentioned the market as the prime inspiration for innovations and industrial development. This is the explanation of the surges of new technology that are created by great wars. The armed forces have not only commanded enormous financial resources, they have also provided science and industry with a highly competent customer. Happily, however, there are other well-qualified customers for new technology. Utilities and the vast public telecommunications and transportation systems have proved themselves capable of inspiring innovations on a large scale. Lately environmental demands and regulations have forced the development of a new type of protective technology concerned with the abatement of pollution and ecological disturbances.

To my mind, the climate of innovations and the existence of an articulated market are the two overriding motivations of industrial creativity. Numerous other prerequisites for innovation have been proposed, although some are debatable. One concerns the availability of venture capital. The imaginative and active businessman has been a partner of the inventor in the success of many innovations. We talk of this as the inventor-entrepreneur relationship. An increasing bureaucracy is stifling the creation of new ventures. Not only has red tape entwined small business, but the many regulations and requirements to protect the public and the environment have greatly increased development and marketing costs. Federal procurement, which used to stimulate new industry, is gradually being fenced in.

On the doorstep of the ninth decade of the twentieth century two new trends may be observed. One is a growing suspicion of a negative relation between new technologies and employment. Unions are sensitive to the introduction of numerical control and robotization in manufacturing. Neither are they convinced that the remaining jobs will be of a sufficiently high quality. Politicians, who always have the impossible task of balancing incompatible demands, will have added difficulties in weighing the support of creative efforts against job security and regional stability. As an engineer I am convinced that the shift of jobs can be accommodated without much hardship to the people involved, given enough time and political common sense. The reemergence of the Luddites of the Industrial Revolution cannot, however, be ruled out completely.

The other tendency is the condition of zero growth in academic science. For a number of years funds for basic research have been curtailed in many countries. In the United States, academic science had reached the zero-growth stage ten years ago. This would not be so serious if most posts in universities and institutes of higher learning were now filled by rather young persons. Retirements will be rare for decades and with no expansion hiring will grind to a halt. This makes the creation of alternative careers for science graduates imperative. If we do not take the necessary steps, we will see an unavoidable atherosclerosis in universities and a drastic setback of scientific accomplishment.

I believe we can discern two primary phases of productivity in science and technology. Creativity is a product of the man and the society around him. Innovation is a product of a man and a market. Creativity needs to be encouraged by cultural forces in society as a whole. Learning, spiritual development, and curiosity are indispensable to it. With good understanding, society should trust its scientists and try to avoid regulation of their activities. Free debate and exchange of ideas can be relied upon to promote originality, but also call upon a society to be receptive to change.

Innovation, on the other hand, depends upon creative organizations. These should scan for new possibilities and carefully watch out for threats. They should keep developmental activities in fruitful contact with markets and cooperate continuously with their customers or clients. They should be conditioned by an appetite for constant progress and refresh themselves by frequent contact with the best universities and research institutes. Their management needs are distinctive and they must organize for flexibility while aiming to recruit the best talent they can find. The creative individual finds his counterpart in such organizations. Can it be enough, however, to prescribe concentration, readiness to learn, and responding to the exceptional wherever it appears? The creative man or woman, let us concede, has chosen his or her parents carefully and, whatever the secret, did not derive it from a formal essay on creativity!

2
Science, Technology, and Economic Growth
Edwin Mansfield

By economic growth economists mean the growth of per capita output. Although there obviously are many social objectives other than the promotion of economic growth, most societies seem to favor it. At the outset it is worth noting that many of the great economists of the past were not very optimistic about how much per capita output could be increased. Malthus and Ricardo were about as pessimistic on this score as they could reasonably get. In their view population would expand whenever the standard of living poked its head above the subsistence level, with the result that more and more people would be working a relatively fixed amount of land. Given the law of diminishing marginal returns, it appeared certain to them that per capita output would have to fall.

With all the advantages of hindsight it is relatively easy to see where Malthus and Ricardo went wrong, at least for industrialized nations. They vastly underestimated the power of technological change, failing to visualize the innovations that were to occur. New techniques and products were destined to offset the dreaded law of diminishing returns, allowing food supplies to keep pace with population increase. Science and technology, the subjects of this book, have been mainsprings of economic growth for 150 years.

Economists of today, although their understanding of the determinants of economic growth is still far from complete, are much more aware of the importance of technological change in the growth process than were the economists of a generation ago. Economic growth appears due largely to increases in productivity which in turn depends on the rate of technological change. This chapter will describe what economists think they know about the relationship between research and development (R & D) and the rate of productivity growth. It will provide some new empirical findings concerning the relationship between basic research and the rate of productivity growth. It will then address some of the reasons for the apparent recent slowdown in the rate of productivity growth in the United States as well as some of the steps recently taken by the federal government to examine its policies toward civilian technology.

RESEARCH AND DEVELOPMENT EXPENDITURES AND PRODUCTIVITY INCREASES

For a variety of purposes, purely analytical and policy oriented, it is important to investigate the relationship between the amount spent by an industry or firm on R & D and its rate of productivity increase. During the past 20 years a number of studies of this kind have been carried out. They are by no means free of problems, in ways I have discussed elsewhere.[1] Perhaps the most important of their findings is that R & D has a very significant effect on the rate of productivity increase in the industries and the time periods that were studied. Minasian found that chemical firms' rate of productivity increase was directly related to their expenditures on R & D.[2] His results indicated that, during the period to which his data pertained, the marginal rate of return—that is, the rate of return from an additional dollar spent—was about 50 percent for R & D in chemicals. A study of my own indicates that the marginal rate of return from R & D was about 40 percent or more in the petroleum industry and about 30 percent in the chemical industry.[3] In agriculture, Griliches found that output was related in a statistically significant way to the amount spent on research and extension. Assuming a six-year lag between research input and its returns, his results indicated a marginal rate of return from agricultural R & D of 53 percent.[4] Another study, by Evenson, used time-series data to estimate the marginal rate of return from agricultural R & D, finding 57 percent.[5] Peterson's study of R & D in poultry indicated a marginal rate of return of about 50 percent,[6] while Schultz's early study of all agricultural R & D indicated a marginal rate of return of 42 percent.[7]

During the 1970s these studies were extended. Griliches attempted to estimate the contribution of public and private research expenditures to the increase in output unexplained by other inputs included in the analysis. He concluded that about two-tenths of one percentage point of the annual rate of growth of United States output was due to R & D expenditures.[8] In a related study he used data for almost 900 manufacturing firms to examine the relationship between R & D and the rate of productivity growth. The results indicated that the amount spent by a firm on R & D was directly related to its rate of productivity growth. Also, he found that the private rate of return from R & D was about 17 percent. It seemed much higher than this in chemicals and petroleum and lower in aircraft and electrical equipment. He found that the returns from R & D seemed to be lower in industries where much R & D is federally funded.[9]

Terleckyj analyzed the effects of R & D expenditures on productivity change in 33 manufacturing and nonmanufacturing industries during 1948-1966. In manufacturing, the results seem to indicate about a 30 percent rate of return from an industry's R & D on its own productivity. In addition, his findings show a very substantial effect of an industry's R & D on productivity growth in other industries, resulting in a social rate of return greatly exceeding that of 30 percent.[10] Nadiri and Bitros constructed an econometric model in which output was treated as an exogenous variable; and R & D, labor, and capital inputs were regarded as functions of input prices, sales, the rate of capacity utilization, and the lagged dependent variables. They concluded that labor productivity is significantly affected in both the short run and the long run by the level of a firm's R & D expenditures.[11]

In interpreting the rates of return obtained, both in these studies and in those done during the 1960s, it is important to distinguish between *private* rates of return and *social* rates of return. The private rate of return is the rate of return to the firm carrying out the R & D; the social rate of return is the rate of return to society as a whole. Since the firm that carries out the R & D frequently cannot appropriate many of the benefits of doing so, the social rate of return may be considerably in excess of the private rate of return. For this reason, many of the rates of return cited above, since they are social rates of return, are likely to be higher than the private rate of return to the firm carrying out the R & D.

How did the results of the studies of the 1970s compare with those of the studies of the 1960s? In general, they are quite consistent, in the sense that they continue to indicate that the level of R & D seems to be closely related to the rate of productivity growth, and that the marginal rate of return from investment in R & D is high, although perhaps not as high as in earlier years. The high marginal social rate of return from R & D is important because it suggests that there may be an underinvestment in R & D, a phenomenon that many economists attribute partly to the differences between private and social rates of return from R & D.

SOCIAL AND PRIVATE RETURNS FROM PARTICULAR INNOVATIONS

The statistical studies just described have been used by economists to estimate the social rates of return from investments in new technology. But they are by no means the only type of study carried out by economists for this purpose. A number of microeconomic studies of the returns from particular innovations have been carried out as well. To estimate the social benefits from an innovation, economists generally have used a model of the following sort. If the innovation results in a shift downward in the supply curve for a product, they have used the area under the product's demand curve between the pre- and post-innovation supply curves as a measure of the social benefit during the relevant time period from the innovation. If all other prices remain constant, this area equals the social value of the additional quantity of the product plus the social value of the resources saved as a consequence of the innovation. Thus, if one compares the stream of R & D (and other) inputs relating to the innovation with the stream of social benefits measured in this way, it is possible to estimate the social rate of return from the investment in the new technology.

To begin with, the studies carried out concerned only agricultural R & D. Although only a few such studies were conducted, notably by Griliches,[12] Peterson,[13] and Schmitz and Seckler,[14] the results were quite consistent in the sense that they all indicated that the rate of return from agricultural R & D in the United States has tended to be high. Until the 1970s, no such estimates were made for industries other than agriculture. In an attempt to help fill this gap, my co-workers and I estimated the rate of return from the investment in 17 industrial innovations, which occurred in a variety of industries and which stemmed from firms of quite different sizes. Most of these innovations are of average or routine importance, not major breakthroughs. Although the sample cannot be regarded as randomly chosen, there is no obvious indication that it is biased toward very profitable innovations—socially or privately—or relatively unprofitable ones.[15]

To estimate social rates of return from the investments in each of these innovations, the model described above was extended to include the pricing behavior of the innovator, the effects on displaced products, and the costs of uncommercialized R & D and of R & D done outside the innovating organization. The results indicate that the median social rate of return from the investment in these innovations was 56 percent, a very high figure. On the other hand, the median private rate of return was 25 percent. In interpreting the latter figure, it is important to note that these are before-tax returns and that innovation is a risky activity.[16]

Also, data were obtained concerning the returns from the innovative activities of one of America's largest firms, from 1960 to 1972.[17] For each year this firm has made an inventory of the technological innovations resulting from its R & D and estimated its effects on its profit stream in detail. When the average rate of return from this firm's total investment in innovative activities was computed the result was 19 percent, which is not too different from the median private rate of return given above. Also, lower bounds were computed for the social rate of return from this firm's investment, which was about double that for its private rate of return.[18]

These findings pertain to the average rate of return. As pointed out in the previous section, econometric investigations by Griliches, Terleckyj, Minasian, myself, and others indicate that the marginal rate of return has also tended to be high. In sum, practically all of the studies carried out to date—both those done before 1970, and those done since then—indicate that the average social rate of return from investments in new technology in both agriculture and industry has tended to be very high. Moreover, the marginal social rate of return also seems high, generally at least 30 percent. As I have stressed in this chapter and elsewhere, there are a variety of very important problems and limitations inherent in each of these studies. Certainly, they are frail reeds on which to base policy conclusions. But recognizing this fact, it nonetheless is remarkable that so many independent studies based on so many types of data result in so consistent a set of conclusions. As noted above, many economists view these conclusions as evidence of an underinvestment in R & D.

BASIC RESEARCH AND PRODUCTIVITY GROWTH

The foregoing studies look at the relationship between total R & D input and productivity change, but they tell us nothing about the effect of the composition of an industry's or firm's R & D on its rate of productivity change. In particular, they tell us nothing about the role of basic research in promoting productivity increase. Basic research is defined by the National Science Foundation as "original investigation for the advancement of scientific knowledge . . . which [does] not have immediate commercial objectives." Does basic research, as contrasted with applied research and development, make a significant contribution to an industry's or firm's rate of technological innovation and productivity change? Although the studies cited above indicate that an industry's or firm's R & D expenditures have been directly related to its rate of productivity change, they have been unable to shed light on this question because no attempt has been made to separate basic research from applied research in development. An econometric study was recently carried out to determine whether an industry's or firm's rate of productivity change in recent years has been related to the amount of basic research it performed, when other relevant variables (such as its rate of expenditure on applied R & D) are held constant. This study has various limitations, but its findings seem of interest, particularly since so little research has been done on this score.

The results indicate that there is a statistically significant and direct relationship between the amount of basic research carried out by an industry or firm and its rate of increase of total factor productivity when its expenditures on applied R & D are held constant. To some extent, this may reflect a tendency for basic research findings to be exploited more fully by the industries and firms that were responsible for them. Or it may reflect a tendency for applied R & D to be more effective when carried out in conjunction with some basic research. Whether the relevant distinction is between basic and applied research is by no means clear: there is some evidence that basic research may be acting to some extent as a proxy for long-term R & D. Holding constant the amount spent on both applied R & D and basic research, an industry's rate of productivity increase seems to be directly and significantly related to the extent to which its R & D is long term. This seems to be the first systematic evidence that the composition, as well as the size, of an industry's or firm's R & D expenditures affect its rate of productivity increase.[19]

CHANGES IN THE COMPOSITION OF INDUSTRIAL R & D EXPENDITURE

If it is true that an industry's rate of productivity increase is affected by the extent to which its R & D is long term, one is led to investigate the extent to which changes have occurred in recent years in the composition of various industries' R & D expenditures. There is a widespread feeling that industry has been devoting a smaller share of its R & D expenditures to basic research, long-term projects, and risky and ambitious projects. Unfortunately, however, little data have been available on this score. To help fill this gap, information was obtained from 119 firms concerning the changes that have occurred in this regard between 1967 and 1977, and the changes they expect between 1977 and 1980. The firms included in the sample, all of which spent over $10 million on R & D in 1976, accounted for about one-half of all industrial R & D expenditures in the United States in 1976.

The results of this survey indicate that the proportion of R & D expenditures devoted to basic research declined between 1967 and 1977 in practically every industry. Moreover, this finding is corroborated by the recent study by Nason, Steger, and Manners.[20] In the aerospace, metals, electrical equipment, office equipment and computer, chemical, drug, and rubber industries, the proportion devoted to basic research dropped substantially. In the sample as a whole, the proportion fell about one-fourth, from 5.6 percent in 1967 to 4.1 percent in 1977. According to the firms' forecasts, there is no evidence that this drop will continue during 1977-1980, but neither is there any evidence that the proportion will rise very much. For the sample as a whole, the proportion is expected to be about 4.3 percent in 1980.

In four-fifths of the industries, based on a rough measure of the perceived riskiness of projects, there was also a decline between 1967 and 1977 in the proportion of R & D expenditures devoted to relatively risky projects. In some industries, like metals, chemicals, aircraft, drugs, and rubber, this reduction has been rather large. Between 1977 and 1980, there will be some increase overall in the proportion devoted to relatively risky projects, according to the firms' forecasts, with the result that the average proportion for the sample as a whole is expected to get closer to its 1967 level.

Despite the decrease in the proportion of R & D expenditures devoted to basic research, the proportion devoted to relatively long-term projects did not decline appreciably between 1967 and 1977 in the sample as a whole. In some industries, like aircraft, chemicals, metals, and rubber, there was a substantial decline, but in other industries, like drugs, there was an increase in this proportion. The proportion of R & D expenditures aimed at entirely new products and processes (rather than improvements and modifications of existing products and processes) declined somewhat between 1967 and 1977. Again, there are marked differences among industries in the amount and direction of change of this proportion.

Why have so many firms cut back on the proportion of their R & D expenditures going for basic research and relatively risky projects? The reason most frequently given by the firms was the increase in government regulations, which has reduced the profitability of such projects. This reason was advanced particularly often by the chemical and drug firms. Another reason advanced by some of the respondents was that breakthroughs are more difficult to achieve than in the past, because their fields have been more thoroughly worked over. In the survey conducted by Nason, Steger, and Manners, the most frequently given reason was that firms have come to view R & D as an activity that can and should be managed in more detail than was earlier thought optimal.[21]

RECENT CHANGES IN THE RATE OF PRODUCTIVITY INCREASE

Finally, let us turn to recent changes in the rate of productivity increase in the United States. It is well known that output per man hour has been growing at a decreasing rate. According to the 1979 annual report of the President's Council of Economic Advisers, output per man hour increased by only 0.4 percent between 1977 and 1978. During 1973 to 1977, its annual rate of increase was 1.0 percent; during 1965 to 1973, it was 2.3 percent; and during 1955 to 1965, it was 3.1 percent.

Economists generally prefer total factor productivity to output per man hour since the former includes nonlabor inputs as well as labor. Total factor productivity is defined by economists as real output per unit of total factor input. Since both labor and nonhuman factor inputs are unadjusted for quality changes, the rate of increase of total factor productivity reflects changes in the average quality of inputs, as well as changes in technology, changes in efficiency, and other factors discussed below. The rate of growth of total factor productivity, like the rate of growth of output per man hour, has declined in recent years in the United States. Whereas total factor productivity increased by 2.5 percent per year between 1948 and 1966, it increased by 1.1 percent per year between 1966 and 1969, and by 2.1 percent per year between 1969 and 1973.

Of course, this slackening of United States productivity growth may be attributable, at least in part, to factors other than a slowdown in the rate of technological change. According to the Bureau of Labor Statistics, one of the major factors responsible for this slowdown has been the increase in the proportion of youths and women in the labor force. Output per man hour tends to be relatively low among new entrants into the labor force and among women. During the late 1960s, new entrants and women increased as a percentage of the labor force. According to the bureau, this change in labor force composition may have been responsible for 0.2-0.3 percentage points of the difference between the average annual rate of productivity increase in 1947-66 and that in 1966-73.

Another factor that has been cited in this regard is the growth of the capital-labor ratio. Between 1948 and 1973, relatively high rates of private investment resulted in an increase in the capital-labor ratio (net nonresidential capital stock divided by aggregate hours worked in the private nonfarm sector) of almost three percent per year. After 1973, relatively low rates of investment have meant that the capital-labor ratio has increased by only about 1.7 percent per year. According to the Council of Economic Advisers, this reduction in the rate of increase of the capital-labor ratio could have reduced productivity growth by up to one-half of a percentage point per year.

Still another factor that is frequently cited in this connection is increased economic and social regulation. In recent years, a variety of new types of environmental, health, and safety regulations have been instituted. Since reduced pollution, enhanced safety, and better health are generally not included in measured output, the allocation of a greater proportion of society's resources to meet these regulations often results in a reduction in measured productivity growth. In addition, it is often asserted that the litigation and uncertainty associated with new regulations discourages efficiency and investment and that the form of the regulations sometimes inhibits socially desirable adaptations by firms. According to the Council of Economic Advisers, the direct costs of compliance with environmental, health, and safety regulations may have reduced the growth of productivity by about 0.4 percentage points per year since 1973.

Although there is general agreement that the foregoing factors are important, many observers feel that these factors are only part of the story, and that the slowdown in productivity growth also reflects a slowdown in the rate of technological change. Thus, some economists have attributed part of the slowdown in productivity growth to the reduction in the rate of increase of intangible capital due to the decrease in the percentage of gross national product (GNP) devoted to R & D during the late 1960s and early 1970s. According to the National Science Foundation's figures, R & D expenditures as a percentage of GNP decreased in the United States from 2.99 percent in 1964 to 2.25 percent in 1976. To my knowledge, no study has been carried out to determine how much of the slowdown in productivity growth could be due to this factor. In part, the lack of such studies is undoubtedly attributable to the great theoretical and empirical problems they would face.

Besides being concerned about the slowdown in our rate of increase of productivity, many observers are concerned about the apparent reduction of America's technological lead over other nations, and with the slow rate of productivity growth in the United States relative to other major countries. According to the Department of Labor, the percentage gain in output per man hour during 1960-1976 was smaller in the United States than in France, West Germany, Japan, or the United Kingdom. During this period, output per man hour grew by about 290 percent in Japan, as contrasted with about 60 percent in the United States. If comparisons are made of rates of increase of total factor productivity rather than of output per man hour, the results are the same. According to Christensen, Cummings, and Jorgensen, the rate of increase of total factor productivity during 1960-1973 was lower in the United States than in any other country included in their analysis (Canada, France, West Germany, Italy, Japan, Korea, Netherlands, United Kingdom).[22] To put such comparisons in perspective, it is important to recognize that other countries' productivity *levels* tend to be lower than in the United States. Nonetheless, comparisons of this sort have caused considerable concern.

Another development contributing to such concern has been the tendency during 1965-1976 for R & D expenditures to increase as a percentage of GNP in some other countries, like Japan, West Germany, and the Soviet Union, while this percentage has been decreasing in the United States. However, although the percentage of GNP devoted to R & D has increased in these countries relative to that in the United States, the United States percentage is still higher than in most other countries, according to the National Science Foundation's 1976 report on *Science Indicators*. The only country (included in the foundation's comparisons) with a higher figure is the Soviet Union. Since the data for the Soviet Union are not comparable with those for the United States (and experts believe they are probably inflated relative to ours), this comparison obviously should be treated with great caution. Further, international comparisons of this sort are clouded by the fact that a considerable portion of the industrial R & D in some major foreign countries is carried out by United States-based multinational firms. For example, it has been estimated that about one-half of the industrial R & D in Canada and about one-seventh of the industrial R & D in the United Kingdom and West Germany were carried out by United States-based firms in the early 1970s. Still further, a nation's rate of productivity growth depends heavily on how effectively it *uses* foreign and domestic technology, and this may not be measured at all well by its ratio of R & D expenditure to GNP.

PUBLIC POLICY TOWARD CIVILIAN TECHNOLOGY

In response to the slowdown in productivity, as well as to the widespread feeling that the United States lead over its foreign competitors in many areas of technology is declining, many suggestions have been made concerning ways in which the federal government might stimulate civilian technology. In 1978 and 1979, the federal government carried out a Domestic Policy Review on Industrial

Innovation. The industry advisory subcommittee involved in this review prepared draft reports on federal procurement, direct support of R & D, environmental, health, and safety regulations, industry structure, economic and trade policy, patents, and information policy. These drafts were discussed and criticized by the academic and public interest subcommittees involved in the review. Further, the labor subcommittee presented a report, as did each of a large number of government agencies. The overall result was a large and far-flung effort to come up with useful and effective policy recommendations.

In the available space, it is impossible to summarize, let alone try to evaluate, the many suggestions that were made. Because of the very tight deadlines imposed on the subcommittees, it would be unfair in any event to criticize the reports for many of their shortcomings. At present, all that I can do is cite two of the general areas that received a considerable amount of attention: regulation and tax credits. One theme that ran through the industry subcommittee's reports was that many aspects of the regulatory process deter innovation. As is well known, there is a strong feeling that this is the case in a number of industries, although it is recognized that we lack very dependable or precise estimates of the effects of particular regulatory rules on the rate of innovation. As might have been expected, the recommendations of the industry subcommittee with regard to regulatory changes were met with considerable opposition, if not hostility, by the labor and public interest subcommittees. Whether many of these recommendations are viable from a political point of view is hard to say.

Some of the industry subcommittee's reports also recommended the enactment of tax credits for research and development expenditures. Perhaps the most important advantage of this mechanism to encourage civilian technology is that it would involve less direct government control than some of the other possible mechanisms. Its most important disadvantages are that it would reward firms for doing R & D that they would have done anyway, that it would not help firms that have no profits, and that it would be likely to encourage the same kind of R & D that is already being done (rather than the more radical and risky work where the shortfall, if it exists, is likely to be greatest). A tax credit for increases in R & D would get around some of these difficulties, but the problem of defining R & D remains. Understandably, the Treasury Department in particular is concerned about this definitional problem.

It is too early to tell what will come out of this Domestic Policy Review. At a minimum, it indicates a laudable interest at the highest levels of government in promoting American civilian technology. Whether it will have much effect depends, of course, on what measures are proposed and whether they are implemented. Neither is known at present. However, one thing is certain: the slowdown in the rate of growth of productivity that occurred during the 1960s and early 1970s is continuing into the late 1970s. And it is having an adverse effect on the rate of inflation, and is being reflected in our rate of economic growth. One can only hope that this Domestic Policy Review will promote the adoption of effective and equitable ways to help reverse this trend without unduly sacrificing other major social objectives. However, given the problems that this exercise faces, and the fact that previous exercises of this sort have not been notable successes, it would probably be unrealistic to set one's hopes too high on this score.

CONCLUSIONS

The conclusions supported by twenty years of economic studies cited above can be stated concisely. Research and development seems to have had a very significant effect on the rate of productivity growth in the industries and time periods that have been studied. The marginal social rate of return from investments in new technology seems to be relatively high, which suggests that there may be some underinvestment in such projects. Holding constant the amount spent on both applied R & D and basic research, an industry's rate of productivity change seems to be directly and significantly related to the extent to which its R & D is long term. During the past decade, firms have tended to cut back the proportion of their R & D going for relatively basic and risky projects. The slowdown in the rate of growth of United States productivity that began during the 1960s and early 1970s is continuing into the late 1970s. In formulating public policy, the federal government should pay much more attention than in the past to the effects of its policies on the rate of technological change in the civilian economy.

NOTES

[1] Edwin Mansfield, "Contribution of R and D to Economic Growth in the United States," *Science* 175 (February 4, 1972); 487-494.

[2] J. Minasian, "Research and Development, Production Functions, and Rates of Return," *American Economic Review* 59 (May 1969): 80-86.

[3] Edwin Mansfield, *Industrial Research and Technological Innovation* (New York: W.W. Norton for the Cowles Foundation for Research in Economics, 1968).

[4] Z. Griliches, "Research Expenditures, Education, and the Aggregate Production Function," *American Economic Review* 54 (December 1964): 961-974.

[5] R. Evenson, "The Contribution of Agricultural Research and Extension to Agricultural Production (Ph.D. diss., University of Chicago, 1968).

[6] W. Peterson, "The Returns to Investment in Agricultural Research in the United States," in *Resource Allocation in Agricultural Research*, ed., W. Peterson (Minneapolis: University of Minnesota Press, 1971).

[7] T. Schultz, *The Economic Organization of Agriculture* (New York: McGraw-Hill, 1953).

[8] Z. Griliches, "Research Expenditures and Growth Accounting," in *Science and Technology in Economic Growth*, ed., B. Williams (London: Macmillan, 1973).

[9] Z. Griliches, "Returns to Research and Development Expenditures in the Private Sector," Conference on Research in Income and Wealth, 1975.

[10]N. Terleckyj, *Effects of R and D on the Productivity Growth of Industries: An Exploratory Study* (Washington, D.C.: National Planning Association, 1974).

[11]M. Nadiri and G. Bitros, "Research and Development Expenditures and Labor Productivity at the Firm Level," Conference on Research in Income and Wealth, 1975.

[12]Z. Griliches, "Research Costs and Social Returns: Hybrid Corn and Related Innovations," *Journal of Political Economy* 66 (October 1958): 419-431.

[13]Peterson, "Returns to Investment in Agricultural Research."

[14]A. Schmitz and D. Seckler, "Mechanized Agriculture and Social Welfare: The Case of the Tomato Harvester," *American Journal of Agricultural Economics* 52 (1970): 569-577.

[15]E. Mansfield, J. Rapoport, A. Romeo, S. Wagner, and G. Beardsley, "Social and Private Rates of Return from Industrial Innovations," *Quarterly Journal of Economics* 91 (May 1977): 221-240.

[16]To extend this sample and replicate our analysis, the National Science Foundation commissioned two studies, one by Robert R. Nathan Associates, and one by Foster Associates. Their results, like ours, indicate that the median social rate of return tends to be very high and much higher than the private rate of return. Based on its sample of 20 innovations, Nathan Associates found the median social rate of return to be 70 percent and the median private rate of return to be 36 percent. Foster Associates, based on a sample of 20 innovations, found the median social rate of return to be 99 percent and the median private rate of return to be 24 percent.

[17]E. Mansfield, J. Rapoport, A Romeo, E. Villani, S. Wagner and F. Husic, *The Production and Application of New Industrial Technology* (New York: W.W. Norton, 1977).

[18]For each year since 1960 this firm has made a careful inventory of the technological innovations arising from its R & D and related activities. Then it has made estimates of the effect of each innovation on its profit stream. In the case of product innovations, the firm calculated for each new product the expected difference in cash flows between the situation with the new product and without it, including the effect of the new product on its profits from displaced products. In the case of process innovations, it calculated the expected difference in cash flow between the situation with the new process and without it, this difference reflecting, of course, the savings associated with the new process. The firm has updated these estimates each year; that is, the firm has revised its estimates of the returns from past innovations. This is of crucial importance, since it means that the firm's estimates for innovations occurring in the early and middle 1960s are based on a decade or more of actual experience, not just forecasts. The data we use are taken from the latest available revision, which was in 1973. Figures are also available concerning the firm's expenditures on R & D and related innovative activities each year. It was by using these cost figures, as well as the figures concerning the total cash flow of benefits stemming from the new products or new processes that came to fruition each year, that we were able to compute the rate of return from the investment that resulted in each year's crop of innovations. It is worthwhile noting that this rate of return is based on the investment in both commercialized and uncommercialized (and successful and unsuccessful) projects.

[19]E. Mansfield, "Basic Research and Productivity Increase in Manufacturing." University of Pennsylvania, 1979.

[20]United States National Science Foundation. *Support of Basic Research by Industry*, eds., H. Nason, J. Steger, and G. Manners (Washington, D.C.: U.S. Government Printing Office, 1978).

[21]*Ibid.*

[22]L. Christensen, D. Cummings, and D. Jorgenson, "An International Comparison of Growth of Productivity, 1947-1973," Conference on Research in Income and Wealth, 1975.

3
Science and Technology in Global Development
Sumitro Djojohadikusumo

ABSOLUTE POVERTY AND THE WORLD'S ECONOMIC PRIORITIES

In 1975 three-quarters of the world population living in the developing countries received a little over one-fifth of total world income. The most serious economic problem in the global context is that under present conditions about 800 million people in the developing countries live in absolute poverty with incomes too low to insure adequate nutrition and without access to essential public services. Two-thirds of the world's absolute poor live in four large countries of Asia: Bangladesh, India, Indonesia, and Pakistan. The growth rate for industrialized economies is estimated at 4.2 percent per year, on average, from 1975 to 1985, while that for developing countries is projected to continue at an average of 5.1 percent per annum. The low-income countries of Asia (less than $250 GNP per person, as defined by the World Bank) may accelerate from their crippling rate of 2.4 percent between 1960 and 1970 to 5.1 percent for the period 1975-1985.[1] Any real increase in the rate of growth of the poorest countries hinges on their hope for improvements in agricultural performance.

Even with growth in income, absolute poverty will continue to be a problem of immense dimensions in the developing countries. Even though the proportion of absolute poor in their total population is expected to decline from 37 percent in 1975 to 17 percent in the year 2000, the number of absolute poor will still amount to 600 million, of which 540 million will be living in the low-income countries.

The eradication of poverty and unemployment must become economic priorities of the first order in a drastic reorientation of development strategies. Such a shift in policy orientation poses daunting problems, as it involves nothing less than an overall transformation of structural relationships in the way resources are accumulated and allocated. The challenge is multiplied by resistances within the social fabric of many developing countries.[2]

As the low-income countries of Asia have predominantly rural economies, policies to alleviate poverty must be directed at agricultural income. In Bangladesh more than 90 percent of the population lives in rural areas, followed by Indonesia with 80 percent, India with 78 percent, and Pakistan with 73 percent. Many of the poor in these areas have no access to land and even farmers depend upon other employment. We might do well to heed a poignant warning voiced by Laughlin Currie. He argues that a concern with poverty ought not to prompt a *direct* attack on the plight of the poor in rural areas. Although he readily agrees that the problem is located there, he maintains that the solution is to be sought in the cities. He explains that the income and price elasticity of demand for farm products is so low that any success in increasing the physical efficiency of large numbers of farmers would "so worsen the terms of trade as to threaten the ruin of all but the most favorably placed." An increase in the income of all farmers at once would make them all worse off in cash income terms. Hence Currie arrives at the conviction that "the solution of the problem of rural poverty must be sought in improving the well-tested economic mechanism of mobility—of aiding in the process of creating more jobs to make things for which there is a high latent demand and a high income inelasticity of demand, and jobs with such characteristics are most efficiently performed in cities."[3]

Much of the public policy discussion regarding inflation betrays a lack of attention to the cardinal importance of stabilizing the incomes of agricultural smallholders. Anti-inflation policies in the developed part of the world have up to now in general relied on a combination of high interest rates and efforts to reduce government spending across the board. These are measures that act at macro magnitudes and contain no element of structural specificity in their operation. But the distributional aspects of the burden of inflation are of urgent importance in the Third World. Maldistribution of incomes precludes demand for supplies on a large scale. The 800 million people living in absolute poverty are just unable to meet the total costs of equipment and fertilizer needed to drive the process of development.

On the demand side for resources, contemporary evidence demonstrates built-in rigidities. Particularly among middle-income and higher income groups there is a basic conservatism in the direction of demand as well as in other patterns of life.[4] As the demographer Nathan Keyfitz has observed, advanced life-styles commit the affluent to continued claims on the world's resources. As income rises, per capita consumption goes up. The movement of people from poorer groups into the middle class will have more impact on the resources of the world than an increase in their raw numbers. He suggests that a middle-class person has at least five times the material impact of a peasant.[5]

Agricultural research must be viewed in this wider context of remedying poverty. When most people talk about plans of integrated rural development, one would think that what is involved is a very fundamental restructuring of rural society. Fundamental reform depends upon people who are task oriented. What I see, however, are mainly groups in governmental and social organizations content to demonstrate their awareness of the problem. What is needed is an attack on rural poverty conducted in terms of struggle for goals. The place to be given to each investment must be specified, including particularly the human investments such as education and health. Note that I am not advocating a return to the simpler living pattern that formerly prevailed in rural communities. Reform is needed because inequalities must be overcome at the same time that new techniques are introduced. Otherwise promising developments such as new high-yield grain varieties will benefit only the stronger, richer groups. Evidence from studies of rice farming in the Philippines and elsewhere in Southeast Asia indicates that small farmers lag behind in adopting new techniques. We need to be sensitive to needs relating to the full living context, such as the overharvesting of firewood prompting a diffuse energy crisis with devastating effects on woodland that Erik Eckholm regards as "the most profound ecological challenge of the late twentieth century."[6] It seems that the best way to help small subsistence farmers is to continue to develop technological innovations, based on plant variety improvements, to broaden our understanding of social conditions in which farmers use them, and to overcome constraints that slow their adoption. What the findings suggest overall is that "to help small farmers in Asia, emphasis needs to be placed on ways to enhance technological change or a more rapid shift in the production and supply function."[7]

DELINEATING A REVISED GLOBAL ECONOMIC SYSTEM

Developing nations need an international framework which is conducive to and can reinforce national growth-directing policies. One reason the industrialized countries have an important stake in the accelerated growth of developing countries is the latter's role as important customers. Over the past 25 years developing countries have emerged as a major market for exports from the industrialized world. In their trade with Western Europe exports of manufactures of industrialized countries increased by 38 percent from 1960 to 1975. The increase in such trade with developing countries was not less than 29 percent during the same period.[8]

The intensifying effort of the Third World to establish a new international economic order is an external projection of the structural changes considered necessary in the patterns of growth and the direction of investments. Needs for continued access to external markets and for stability in export earnings will be continuously stressed. Governments of the developing world have been incessantly exhorted to get away from import substitution and rely instead on export orientation. However, where governments have applied this gospel and seek to reap the benefits of export development, they are faced with tariff and nontariff barriers.

The conditions and trends they have encountered prompt developing countries to put forward demands for organized trading served by long-term commodity agreements combined with buffer-stock financing. The tedious negotiations underway on the integrated program for commodities and buffer-stock financing are directly connected with the stake that the developing countries have in continued access to external markets and stability in export earnings at an expanding rate. The results on both fronts have been meager to date. Spokesmen from the North, especially from the EEC, commend to the South the system of "compensatory financing" employed by the IMF and the STABEX system outlined in the LOME agreement between the EEC and the A.P.C. group of developing countries. While these arrangements are helpful to a group of countries, they tend to introduce further distortions because they exclude others.

Agglomerations of Economic Power and the Political Economy of Trade

A striking feature of the global scene is that its economic activities appear to evolve from and around agglomerations of economic power such as the EEC, the United States, Japan, and the COMECON, with China emerging as a major factor. Transcending the delineations of regional entities and the boundaries of nation states are the operations of the transnational corporations. Furthermore, in connection with the problems of materials resources, a delineation or at least a distinction can be made between resource-rich and resource-poor or rather resource-owning and resource-searching countries. The latter kind of grouping will sometimes cut across the boundaries of regional and subregional spheres and at other times coincide with some of them.[9]

Agglomerations of economic power act as "poles of concentration" within which and around which major and dynamic economic activities rotate. They represent economic entities of such importance that their performance will largely determine the volume of world trade. They represent large and cohesive units within which international political or economic pressures can work to mold the commercial policies of bigger segments of the world trading system. They are geopolitical facts exerting a political impact by their inactions as well as by what they actively carry out. The trading policies and practices displayed by these powerful economic entities so far indicate trends inimical to world trade expansion.

The most notable of these trends is the expansion of transnational corporations. They are dynamic in nature and operate everywhere in the world. Their practices challenge public policies in the developing world, just as they do in the industrialized nations. Their intracorporate procurements and locational decisions have replaced the market and other traditional determinants of the level and pattern of trade. They operate with decision rules that transcend even the frameworks of supranational blocs and groupings. While transnational corporations are facts of life, neither good nor bad in themselves, they operate in the environment of nation states and will have to be keenly aware of their needs. One of the most important functions they can perform is to act as linkage institutions between the "traditional" sector of developing countries and their more modern, technologically based segments.

Former Secretary of State Henry Kissinger has referred to transnational corporations as "engines of growth" that would benefit the progress of developing societies through the deployment of their store of wealth in terms of technology. Admittedly, the transnational corporations do excel in two fields. They have accumulated vast technological and marketing knowledge to which they are constantly adding by their R & D efforts and they have developed a highly effective process of transnational decision making. In fact the dynamics of their research leading to new technology and process knowledge is almost the defining characteristic of the modern transnational corporation. This large and rapidly expanding stock of technical and production knowledge is of enormous potential benefit to men and societies. In many instances the transnationals' impact does' further human welfare. But it must be stressed that there is.nothing in the inherent logic of their behavior which assures that a transnational's actions will be appropriate to given situations in any country. I emphasize here that I am talking about the logic of the inherent nature of transnationals, not the motives, whether for good or evil, of their executives.

Unless channeled toward or reconciled with public policy goals of nation-states the use of their capacity may well increasingly clash with the larger social dimensions of the emerging development trends. The development fostered by transnationals in our part of the world, while making many important contributions, has not been fully responsive to social needs, particularly those of the lower income groups of society. The motive power of transnational corporations is their need to expand and grow continually. Therefore they must have an increasing number of responsive buyers, which has led them to develop extensive links in the affluent strata of society. Therefore, while they may be engines of growth the question still remains: For whom? Their activities are not geared per se to those of development, and in the absence of proper public policy, transnational corporations could tend to accentuate rather than reduce income disparities in poor societies.

The Dynamics of Technology Transfer

Understanding the role of the transnationals has become necessary in order to assess the issues related to the transfer of technology. As already noted, transnationals are the major sources of new production and technical knowledge. The research results of transnational corporations are often supplied by these private enterprises but this knowledge, and therefore any scientific build-up based on it, is generally carefully guarded and subject to their own imposed restrictions. When we speak, too glibly, about "transfer" of technology, we must remember that technology is closely related to the dynamic fields of science and engineering, continually responding to challenges and building on themselves. Particular artifacts or results of a dynamic process may be transferred, but not the process itself. Process knowledge and designs are typically proprietary items for transnationals and one pays very dearly through license fees or royalties for the right to use this knowledge, and even then one's rights may be sharply circumscribed by restrictive covenants in the license agreement. Thus one does not buy or borrow or transfer the whole of the technology let alone the science as such. These have to be developed to be fully owned, to be fully at one's own disposal.

Thus one can accept the appropriateness of transferred technology only to the extent that this is regarded as comparable to the role of bridge financing in the world of finance. In financing a project or an industrial plant, banks provide bridging finance in order to utilize the bridging time thus provided to organize and consolidate the project's financial liquidity by arranging for the long-term finance which is the fundamental requirement of the financial structure of a project or program. Of course, all comparisons suffer from deficiency, but this is what I think of when people talk about transfer or buying technology. One buys only a partial right to use technology and that can be characterized as transferring technology provided it is understood as a bridging phase in the course of a program to develop one's own capabilities in a longer time dimension, such as a five- to ten-year bridging phase in a ten- to twenty-five-year process of developing one's own capabilities. Unless that consolidation does take place in the form of developing the indigenous basic science and applied sciences and the indigenous technology over that period of ten or twenty-five years, then there is no chance really of fully developing one's own potentialities. Unfortunately many experts of the developing countries themselves are not always sufficiently sensitive to this range of issues.

The constellation of forces affecting technology policy and investment decisions is by no means neutral. For a variety of reasons, these factors often favor the transfer of unadapted capital-intensive technology. This is the case for large-scale public sector projects included in development plans or sponsored by foreign aid programs, in private investments by transnationals, and in activities supported by financial institutions under their influence. Governments can play a role by diverting incremental resources to activities which reinforce indigenous capabilities in science and technology and gear other aspects of projects to the needs of their societies. Here is where the time dimension needs to enter policy thinking.

The Changing Character of Trade Relations

For the growth and exports of both agricultural commodities and manufactures the developing countries are highly dependent on the development of world markets and thus on the policies of the industrial countries. Here the world picture for the developing countries is presently characterized by features which continue to underline their trade positions. The secular decline in their terms of trade has been persistently accompanied by short-term cyclical fluctuations which prevent readjustments serving longer term purposes. This has been greatly aggravated by discriminatory restrictions imposed by the industrialized countries on processed agricultural commodities and on manufactures from the Third World.

Trading procedures and arrangements have not yet been fully adjusted to the new forces shaping world trade. The present unresolved crisis in international monetary and exchange arrangements is simply the financial analogue of this more general situation. The expansion of world trade was led by trading activities primarily among industrialized countries, boosted by regional integration in Western Europe. The 1960s may be considered a stable period in world trade, but the share of the Third World in total world exports then fell from 30 percent to 20 percent. From 1950 to 1970 the Third World's share of primary product export markets fell from 52 percent to 44 percent.

Our terms of trade have been deteriorating steadily. Where export volumes have been increased, changes in the terms of trade have eroded the import purchasing power of our earnings. Prices of agricultural commodities, particularly agricultural raw materials, declined through the 1950s and early 1960s relative to the prices of manufactured exports of industrial countries. During that period the prices of minerals and metals have been subject to wide fluctuations and so far with no clear trends. The increase in oil prices since 1973 produced of course a sharp improvement in the terms of trade for net exporters of oil but it worsened the terms of trade for all other developing countries. The deterioration was particularly severe for the low-income countries, which further limited the benefits they derived from the expansion of world trade.

A proposal put forward by some German academic economists may exemplify the structural approaches that are needed. What they suggest is a change from a "complementary" to a "substitutive" division of labor. The focus of activity would shift under the latter concept from an industrial country where R & D are carried out for a product to first-generation manufacture in another, less developed country, followed by a shift of production to countries of low labor costs in the Third World.[10] This proposal does delineate an area of opportunity for developing countries. But it deals with only part of the spectrum, ignoring mineral processing, forest products, and other agricultural commodities. Secondly, such a proposal could not be adopted unless many of the other items on the North-South trade agenda were resolved first, with specific reference to markets and the transfer of technology. The present reaction in the United States, the EEC, Australia, and even Japan to imports from developing countries must give us pause.

In coming decades the political economy of international relations will center on the role and relative importance of material resources. Considerations of resource policy and resource management will largely determine the behavior of nation-states on the domestic level and in external affairs. In a world of resource imbalances, the countries owning resources will seek the maximum benefits from their good fortune. They will insist on higher economic rents and on sharing in monopolistic rents. They will press demands for in situ processing of the basic material (whether in ores or timber) and subsequently for forward integration into the fabricating process. They will press for earlier and earlier integration into the phases of the production cycle.

Primary commodity-producing countries are moving into the processing stages and this tendency to go "downstream" can only be expected to accelerate. This shift in the global distribution of processing, converting, and fabricating of raw materials will have several consequences of far greater importance than might at first have been thought. From the point of view of intermediate industries in industrial countries, fewer and fewer raw materials will be coming on to the world market and with stable or growing demand the long-term trend of raw materials prices can be expected to rise. To the extent that increased consumption of the processed raw material takes place within the primary producer, which is likely to be the case with rising incomes and as industrialization takes hold, the amount of the processed materials going on to the market will also be less than would otherwise be the case. This would tend to reinforce the secular upward bias in international trade prices. Finally, in an integrated production process monopoly rents are more easily captured at intermediate or final stages than in raw material production. Thus the move "downstream" will serve the developing countries' aim of capturing more of these rents as well as increasing the value added within their own economies.

DEVELOPMENT OF A SCIENCE AND RESEARCH COMMUNITY

The governments of developing countries must develop their own experts, research scientists, and engineers, in order to be able to screen proposals and confine capital-intensive technologies to uses where they are clearly superior to the alternatives. More rigorous application of social benefit analysis needs to be made by planning authorities to be sure that the utilization of total resources is enhanced, not just the productivity of labor. The research capability of developing countries must be greatly increased through the establishment and expansion of high-quality research institutes based on the build-up and development of fully qualified research manpower. Although the process of developing indigenous capacity for scientific research is a long-term process, it must be started immediately and must be accorded high priority in current development plans as it is of real and fundamental importance. Failing this, countries depending on imported, bought, or transferred technology will always be running several decades behind the technologically active countries.

Unless a country is able through its own research and scientific activities to achieve an awareness of significant current developments of world science, that country is unlikely to reach a threshold of technical effectiveness. It is only through such a process that a society and its research community will be able to select and negotiate the purchase and ensure the effective assimilation, and therefore the adaptation, of what is appropriate among the alternative technologies as required by its economic and social goals.

I cannot stress enough how important it is that high-level manpower training programs be formulated and implemented in a concentrated way so as to develop both scientists and high-quality research institutes. It is important that there are local institutes that can focus on problems of great importance to the societies concerned; otherwise important work may be left undone for want of a place in the research priorities of rich countries.

After considering the relationship among fundamental research, applied research, technology, and development, it appears to me from personal observations and many discussions that in the developing world today fundamental research is in danger of being neglected in favor of applied activities. Modern science has been directed to the dual function of gaining greater knowledge of natural phenomena and to harnessing that knowledge to meet man's needs. The process is an interactive one, requiring all its different phases. It is often the needs stemming from development and the improvement of techniques that induce stepped-up efforts within pure science. It can be harnessed to the tasks of development in order to yield solutions to the serious problems which now confront us.

The creation of a strong science and research community enabling a fruitful transfer of technology from other countries with different cultural backgrounds is highly dependent on education and the educational system of the society under consideration. In this context an interesting point is raised by F. Bary Malik of Indiana University. In the search for factors common to societies progressing to a better living standard through the application of technology Malik has highlighted enrollment in secondary schools. The creation of institutions of higher learning and research is a necessary criterion enabling such a transfer of technology, but it is not yet a sufficient condition. Similarly, primary education is also a necessary criterion but not a sufficient one because the main emphasis at the primary level of education is to teach the basic three R's of a particular culture to children. Hence many countries having almost 100 percent of their young population going to primary educational institutions are still poor—measured in terms of the per capita income or industrial outputs.[11]

Advanced, Adaptive, and Protective Technologies

In view of the need for economic development and social progress of societies in developing countries, it is a fallacy to assume that they can do without some of the latest advances in technology. Developing nations need to deepen their capability, knowledge, and understanding of advanced science and scientific research and to adopt advanced technology at least in key areas which are leverage sectors to increase the productivity of the social system as a whole. For material resources, for example, there is the need to develop an advance in theoretical physics, analytical chemistry, biology, biophysics, biochemistry, geology, geophysics, and geochemistry—not to speak of metallurgy and mineral technology. They even need to develop their capability and the application of space technology to assess land and water, vegetation and aquatic resources. From this depth of understanding we can derive ways to adapt the technology of advanced countries to the conditions prevailing in the Third World. That is why I prefer to speak of *adaptive technology* instead of appropriate technology.

(The discussions and contributions so far on what are described as "appropriate" and "intermediate" technology have suffered from excessive generalizations and failure to go deeply into the specifics of problems and possibilities. It would be useful to disaggregate the process of manufacture of a number of products and determine which specific steps are amenable to the use of these alternate procedures and what difference it would make to the quality of the products.)

We must be prepared to emphasize the protective dimension of adaptive technology in order to conserve natural resources and to restore and regenerate to the extent feasible those resources already depleted. Particularly at this stage of global development the relevance of protective technology needs to be emphasized consistently for the present and for the long run. We must realize that some of the most urgent problems include components that can be analyzed successfully only over a longer term, and that even the simplest immediate steps may have long-range implications. While our research must become policy oriented, the longer range questions must not be sacrificed.

Growth and the Environment

The time dimension confronts us continually. The Report of the Workshop on Alternative Energy Strategies alerts us that "the supply of oil will fail to meet increasing demand before the year 2000, even if energy prices rise 50 percent above current levels in real terms."[12] Even if there are elements of speculation in this forecast we have nonetheless entered the stage of relative materials imbalances, with specific reference to energy fuels and certain basic minerals. Under the prevailing conditions of primary reserves these imbalances represent excess effective demand relative to known technology, production capacity, and the circumstances of political economy, including environmental considerations. This situation has further aggravated the plight of the world's poor, magnifying the ramifications of the present maldistribution of world incomes.

It is imperative that growth and development policies emphasize environmental considerations.[13] Environmental impact assessment for all major development proposals would be only a beginning. It has been alleged that economic systems aim to optimize gains over the short term, while ecological considerations suggest ways to minimize liabilities over the long run. If such views ever had validity they lack it now. Much of what was formerly considered social expenditure for environmental conservation must be included in the cost of investment. The concept of cost-benefit analysis must be extended so as to incorporate relevant and appropriate measures for environmental protection as inherent components of cost.

Productivity Centers, Scientific Services, and Engineering

The development of industrial technology entails the reinforcement of small industries related to rural development by centers that specialize in R & D, extension, and low-cost training services. Some of these must also be situated in urban areas. An adequate infrastructure is needed for the full effectiveness of science and engineering communities. We should emphasize manpower development, laboratories, and instruments for servicing activities in support of scientific research, including calibration and systems of standards. Laboratories for the testing of construction materials and other linkages are needed between small- and intermediate-scale industries and major investment projects. Funding for demonstration projects, training personnel from user industries, and communication to the society at large must be provided.

Empirical observations of developing countries bear out the importance of developing local and national engineering firms. As they build up their capability in design engineering they become important dissemination agencies. Through such firms and their linkages, prototypes can be converted into tested, refined machines, and such firms will have a commercial interest in the acceptance of such technologies by users. Centers for documentation and information on scientific research results, clearinghouses, instructional and self-help manuals, trade journals, and newsletters from employers' associations and extension services for small industries should all be coordinated as a comprehensive national technical program.

An exhaustive inventory of natural resources should be undertaken to contribute to the context of decision making about technical programs of all kinds. It should comprise land, soil conditions, water resources, vegetation (forests in particular), aquatic resources, and energy fuels and minerals. To return for a moment to the energy shortage that is stripping forested areas of firewood, solar energy must be developed for cooking and heating and bioconversion must be favored in order to generate gaseous fuel and organic manure from cattle dung and night soil.[14] It is plain that requirements for food and habitation can only be met by imaginative management of land and water resources. Water resources must be seen in direct connection with forestry, as deforestation is responsible for soil erosion on an extended scale and perennial floods. Soil erosion is responsible for the loss of a great deal of potential crop production, declines in soil fertility, increased flooding, and silting of lowlands, irrigation canals, and reservoirs.

Powerful nations and weak ones, the resource rich and the resource poor, ceaselessly espouse "rational approaches" to resource policies. Without fail, each gives a different interpretation of what such rationality must entail. Let us therefore collaborate in an encompassing international effort to inventory our resources and evaluate them as impartially as we can. The advanced technologies developed by the industrial societies can be of invaluable assistance to governments of developing countries in their efforts to ascertain their resource potential on a more factual basis, provided always that such assistance is designed to develop and improve the national capabilities of developing countries to conduct and extend their own resource inventory. In such a context, the results of cooperative endeavors, reinforced and supplemented by regional arrangements, will generate a shared knowledge and enhance a better understanding of the world's resource potential. The developing and the developed nations both stand to gain. This in turn may contribute to some workable consensus as to the desired rationality in resource policy. Appropriate modalities for collaborative arrangements in the resource inventory could mitigate the acrimonious postures of confrontation between resource-owning and resource-searching countries. Hopefully, in the little time left to mankind, inclinations toward confrontation, ever latent where resources are concerned, may give way to accommodation on mutually acceptable terms of equity.

NOTES

[1] International Bank for Reconstruction and Development, *World Development Report 1978* (Washington, D.C.: World Bank, 1979), especially table 34, table 12, pp. 13 and 21.

[2] Hollis Chenery and Moises Syrquin, *Patterns of Development 1959-1970* (Oxford: Oxford University Press for the World Bank, 1975), especially pp. 6-10.

[3] Laughlin Currie, "The Objectives of Development," in *World Development* 6 (January 1978): 1-10.

[4] Gunnar Myrdal, "Environment and Economic Growth," International Conference on Environment, Tokyo (organized by Nihon Kaisha Shimbun), May 1976.

[5] Nathan Keyfitz, "World Resources and the World Middle Class," *Scientific American*, 234 (July 1976).

[6]Erik P. Eckholm, *Losing Ground* (New York: W.W. Norton, 1976), especially pp. 101-113.

[7]Robert W. Heard and Randolph W. Barker, "Small Farmers and Changing Rice Technology," *Ekonomi dan Kavangan Indonesia* 25 (June 1977).

[8]International Bank for Reconstruction and Development, *World Development Report*.

[9]Sumitro Djojohadikusumo, *Science, Resources and Development* (Jakarta: LP3ES, 1977).

[10]A.O. Donges and B. Juergens, "Problems of a New International Economic Order," *Economics* 15 (Institute for Scientific Cooperation, Tübingen, Federal Republic of Germany).

[11]Bary F. Malik, *Transfer of Science and Technology: Prelude to a Perspective* (Bloomington: Indiana University, Department of Physics, 1978).

[12]L Carroll Wilson, *Energy: Global Prospects 1985-2000* (New York: McGraw-Hill, 1977).

[13]International Council of Scientific Unions, *Environmental Impact Assessment: Principles and Procedures* (Toronto: "SCOPE" Report 5, 1975).

[14]United States National Research Council, *Methane Generation from Human, Animal, and Agricultural Wastes* (Washington, D.C.: National Academy of Sciences, 1977).

4
Energy and Civilization
George Basalla

Energy conservation has found critics who claim that if we use less energy we will become less civilized. They believe that the retreat from high energy consumption will lead mankind directly back to the caves which provided his first shelter. Such was the message of an advertisement appearing in *Newsweek* in 1978. Above a picture of a caveman chipping away at a large rock were these words: "If you liked the Stone Age, you're going to love the day oil runs out." The immediate response to this warning is that the first substantial use of petroleum dates not to the Stone Age but to the period of the Civil War. And furthermore, that even a fifty percent reduction in our total consumption of energy would transport us back not to Paleolithic times, nor even to the Dark Ages, but to the 1950s.

Nevertheless, apocalyptic visions of an American society forced to reduce its energy consumption continue to haunt us. We foresee a doomed civilization with tractors paralyzed in the fields, abandoned automobiles rusting on weed-choked freeways, factories as quiet as tombs, and our haggard descendants facing a life of everlasting drudgery.[1] These are strange visions for a nation in which one can use a 5,000-pound automobile to drive a few blocks in order to buy a half-dozen cans of beer that will be drunk in an overcooled room and the empties thrown on the trash heap instead of being delivered to an aluminum recycling center.[2]

How did it come about that people who use and waste vast amounts of energy believe that the alternative to high energy consumption is the primitive condition endured by early man? Even if the choice is not between the Stone Age and life in the twentieth century, what is the relationship between energy use and the level of civilization? In seeking answers to these questions it will be necessary to explore the ideological premises of energy use. Most current commentary on this subject emphasizes resources, economic factors, and political choices, but these do not exhaust the substance of the problem.

THE ENERGY-CIVILIZATION EQUATION

Current approaches to the energy problem ignore the ideological component that has for two centuries pervaded energy consumption in the West. High energy consumption is associated not only with physical comfort, economic well-being, and military strength, but has been identified with civilization itself. The tendency of Western societies to equate energy use with the level of civilization was satirically noted by Aldous Huxley. "Because we use a hundred and ten times as much coal as our ancestors, we believe ourselves a hundred and ten times better, intellectually, morally, and spiritually."[3]

When energy consumption comes to serve as a measure of the level of civilization attained by a nation, changes in energy use will have wide implications. A retreat from rising energy consumption under those circumstances means far more than the minor discomfort of living in a warmer house in the summer and a cooler one in the winter, or driving a smaller car less frequently and more slowly. As less energy is available per capita the nation is thought to lose its standing among the world's civilizations.

Those countries with high rates of energy consumption are ideologically committed to maintaining them and those with lower rates are motivated to copy their energy-hungry, civilized superiors. This ideological component helps to explain why so many of the less industrialized nations felt it necessary to have their own nuclear reactors. It was not necessity that drove them to acquire them but the feeling that they might be left behind in the race toward civilization. Let us postulate an equation between these terms that may be expressed as follows:

$$\text{civilization} = k \text{ (energy)}.$$

Although no one has ever formally written out such an equation it has pervaded Western thought for the past two centuries. It can be found in the physical, life, and social sciences, and in technology, philosophy, and popular culture.

The right side of the equation contains energy, a well-defined physical concept. On the opposite side appears civilization, which is a subjective evaluation of the intellectual, moral, and aesthetic accomplishments of a society. The two sides of the equation are directly related so that high energy consumption results in high civilization and low energy consumption in a low level of civilization. If the use of energy is very low then the society may be placed in the savage or barbarian state that precedes civilization.

Having introduced the energy-civilization equation the two questions raised previously can now be related to it. First, a study of the origins and development of the energy-civilization equation will help to explain why reduced energy consumption today evokes nightmare visions of an uncivilized existence in the near future. Second, an inquiry into the validity of the equation will raise fundamental questions about a formula that brings energy and civilization into so close and precise a relationship.

TECHNOLOGY AND THE EQUATION

The energy-civilization equation originated in the early nineteenth century. Prior to that time the introduction of new energy sources was not linked to the advancement of culture. A case in point is the Middle Ages which witnessed a great power revolution. The waterwheel, windmill, and effective harnesses for draught animals were all first extensively used in the West during the medieval period.[4] Although these new power sources transformed social and economic life no medieval thinker was ever moved to claim that they were the ultimate sources of the cultural and spiritual achievements of the time.

The rose window of Chartres, the philosophy of St. Thomas Aquinas, or any other of the accomplishments of the age were never related to the energy of the water, air, and beasts that had recently been put to new and practical uses. And conversely, medieval man never feared that dry streams, windless days, and bad harnesses would mark the end of his civilization. Yet, by the 1800s the energy-civilization equation found easy acceptance and claims and warnings of this sort were gaining in popularity.

The formulation of the energy-civilization equation was stimulated by the scientific revolution of the seventeenth century. The emergence of modern science, and the subsequent identification of scientific and technical advancement with human progress provided the kind of intellectual environment in which a newly introduced power source would be dealt with differently than it had been in the Middle Ages. The scientific revolution created a world view in which energy and civilization could be directly related.

In the seventeenth century Sir Francis Bacon listed the great inventions that had changed the course of civilization; they were the compass, gunpowder, and the printing press. With the rapid growth of science and technology it became an easy matter to extend Bacon's original list by adding new inventions. By the late eighteenth century an obvious addition to that list was the steam engine, a mechanically ingenious device that produced large amounts of power and had noticeable social and economic effects.

The steam engine was by no means the sole prime mover of the Industrial Revolution. However, it quickly became the symbol of industrialization and the social, economic, and cultural changes that accompanied it. As an example of the civilizing role assigned to the steam engine in nineteenth-century thought consider several typical contemporary responses to it. These responses are chosen from technical treatises written for engineers as well as popular accounts of inventions read by laymen. What these sources share in common is the firm belief that the steam engine is much more than a new source of power. It is seen as the driving force of human progress and betterment and ultimately as the creator of civilized society.

In Great Britain, where the steam engine first appeared, it was admired as a contrivance that simultaneously brought wealth and civilization to the British people. The enthusiasm of those who commented upon steam-powered civilization went beyond simple praise. They felt it necessary to explain in detail how the energy of the coal in the steam engine could be transformed into civilization. Their explanation may be summarized as follows. The burning coal produces the steam power that supplements and surpasses human power. This excess steam power is used to increase the productivity of labor and to create the additional wealth needed to maintain a leisured, educated class. The abundance of steam power also stimulates the growth of the middle class by supplying it with every sort of material goods.

Steam power, which supports the nonproductive educated class and meets the material demands of the consuming middle class, also has the ability to raise the intellectual level of those at the bottom of the social scale. In an age when both education and material goods are being increased by steam power, the habits, manners, and feelings of even the lowliest laborer cannot help but be improved. Therefore, the steam engine not only weaves cotton, powers locomotives, shapes iron, and pumps water. It also influences the moral, aesthetic, and intellectual life of the British people and in that way lifts the entire nation to a higher stage of civilization.[5]

A similar viewpoint was put forth on a more popular level in 1868 by British author John Timbs. According to Timbs, just as man is the noblest work of God so the steam engine is the noblest work of man. That is, man is to God as the steam engine is to man.[6] And why does the steam engine deserve such high acclaim? The answer is inevitable: because it is responsible for the "physical, intellectual, and moral advancement of mankind" in the nineteenth century. The only note of regret to be found in this eulogy to the civilizing power of the steam engine occurred when the author recalled the greatness of Greek civilization. If only the Greeks had had the steam engine, he lamented, just imagine how far even Greek civilization could have been advanced by it. Responding in the spirit of this nineteenth-century writer one might attempt to imagine a steam-powered Platonic dialogue or one of the plays of Sophocles or Aristophanes improved by the addition of the surplus power made available by James Watt's invention. Such a possibility appears ridiculous to the modern reader but less so to nineteenth-century men who were convinced that steam was the force behind their advanced culture and that any civilization could be improved by the addition of new energy sources.

The foregoing examples, drawn from British technical and popular literature, could be readily duplicated in the literature of the other European countries that soon followed Britain's footsteps along the path towards industrialization. But rather than pursue the Continental responses let us turn to the steam engine in America. In 1876 our nation celebrated its centenary at an exhibition held in Philadelphia. The political freedom and growth of the American republic were being celebrated there with political oratory and lavish displays of agricultural and manufactured goods. Overshadowing all the speeches and goods was a huge 1,400-horsepower Corliss steam engine that stood over 40 feet above its platform and supplied power to all of the machinery on exhibit.

The centennial festivities began when President Grant opened the valve of the engine and were not halted until the engine's gigantic flywheel was stopped.[7] Visitors to the exhibition looked upon this steam monster as both symbol and tangible proof of the progress America had made in the civilized world since the signing of the Declaration of Independence. To this day the Corliss engine remains the single best known feature of an exhibition that was mounted to commemorate the first one hundred years of American independence. A steam engine holds this distinction because it was the physical embodiment of the idea that a people who have access to large quantities of energy are progressive, civilized, and superior to those who use less energy.

Up to this point the illustrations chosen have reflected the optimistic side of the energy-civilization equation, the side that correlates rising energy use with higher levels of civilization. There is, of course, a corresponding pessimistic interpretation of the equation that emphasizes falling rates of energy use and the decay of civilization. If, for example, the coal supply should be exhausted then steam-powered civilization must come to a halt. A nineteenth-century American writer who considered this possibility rejected it because he could not imagine that God would not have had the foresight to provide mankind with enough fuel to carry out its appointed tasks on earth.[8]

Not all nineteenth-century thinkers were as naive, or pious, as this author. It had occurred to some of them that there might well not be sufficient energy to maintain current levels of civilized life. One such pessimist was the English economist and philosopher W. Stanley Jevons. In his book entitled *The Coal Question* he warned his countrymen of a coming energy crisis that would threaten Britain's economy and civilization.[9]

Jevons' analysis of his nation's energy resources led him to conclude that its coal supplies were rapidly diminishing. Since he could not conceive of any alternative to coal, Jevons concluded that within a century Great Britain would be forced into moral and intellectual decline. He envisioned some time in the future when a New Zealander would visit the ruins of a coal-exhausted London in order to contemplate the remnants of British civilization just as British tourists were then visiting Rome to mourn its lost grandeur.

Once coal was no longer available to energize the thrust of British civilization the nation would face a melancholy choice. It could use the remaining coal supply to fuel one brief brilliant burst of cultural activity or ration it over a longer period of time while sinking into prolonged mediocrity. Jevons' only hope was that when coal lit other flames of civilization elsewhere, the world would not forget what Britain had accomplished in its coal-rich days. His pessimistic forecast was widely debated in the press and in Parliament. The prime minister argued that steps should be taken to reduce the national debt in light of Jevons' dire predictions. But then as now, the public preferred to believe that something would come to the rescue. New coal supplies were discovered in England and the petroleum industry was born.

SCIENCE AND THE EQUATION

The steam engine appeared to offer strong evidence that energy could be converted into civilization. But neither the steam engine nor any other mechanical device could provide a proper theoretical basis for the energy-civilization equation. Only the sciences could provide such a foundation, and did so during the nineteenth century. First physics and chemistry, then the biological, social, and behavioral sciences were called upon to offer theoretical justification for linking energy with civilization.

Among the great successes of nineteenth-century physics were the discoveries of the laws of conservation of energy and the establishment of a science of thermodynamics. Early in the century scientists had their first glimpse into the possibility of energy conversion. At that time they were interested in the conversions of heat to light, light to chemical action, chemical action to motion, motion to electricity, electricity to magnetism, and so forth.[10] Some men of science were not satisfied to confine the conversion series to the boundaries of the physics and chemistry laboratory. Is it not possible, they argued, to convert physical forces or energies into biological ones? After all it occurs naturally every time an animal assimilates its food. And cannot the series be extended from the biological to the nervous forces that energize the nervous system? And why stop there? Why not take the next step, the one that connects nervous forces to the mind and to the study of moral and intellectual energy?

If one answered "yes" to these questions there existed no theoretical barrier between the physical concept of energy and the moral and intellectual progress that characterized civilization. It was in this way that the energy-civilization equation finally found an apparently respectable place for itself among the sciences. Once the path had been opened between physical energy and culture it was possible to imagine energy conversion sequences that began in the firebox of a steam engine, or in the windings of an electric dynamo, and ended in the world of morality, social and intellectual concern, and artistic creation. Most scientists preferred to work on the first few links of the sequence, links that were clearly empirical. On the other hand, there were those who saw no problem in speaking of vital, mental, and social energies and in determining their relationship to the energy of the physical sciences.

One of those who was prepared to move from physics to culture was the American scientific journalist and publisher Edward L. Youmans. Youmans claimed that the transformation of water, steam, and electrical power into social activities was the greatest accomplishment of the century. He went on to predict that in the future any question dealing with man and society would best be studied in terms of the physical principle of the conservation of energy.[11] Youmans' proposal to study social, economic, and political problems in terms of energy exchanges and conservation gained in popularity during the twentieth century. Let us consider three strikingly different figures who in the early twentieth century attempted to understand society in terms of energy. The first was a distinguished German chemist; the second, a well-known American literary personage; the third, a British chemist. All three helped to bring the broader implications of the energy-civilization equation to a new scientific and popular audience.

The German chemist was Wilhelm Ostwald, winner of the 1919 Nobel prize for his work in physical chemistry. Ostwald was part of an influential group of chemists and physicists known as the energeticists. These men were convinced that the atomic theory of matter should be replaced by an energetic theory of matter. Atoms, they said, were mere fictional entities, products of the scientists' fertile imagination, while energy was a well-defined concept resting upon a firm experimental basis. At the same time that he was engaged in his chemical labors and philosophical speculations Ostwald became fascinated with all aspects of the idea of energy. He renamed his house *energy villa* and a factory he owned the *energy works*. His stationery crest carried the initial E, for energy, and he announced the discovery of an energy-based moral principle which should henceforth govern all of man's actions. It was called Ostwald's energetic imperative and its message was simple: DO NOT WASTE ENERGY!derived[12]

The energetic imperative had as its corollary our familiar energy-civilization equation. But Ostwald modified it somewhat. Not only did the civilized nations of the world have more energy available to them but they used their energy allotments more efficiently than did the uncivilized societies. If one accepted the energetic imperative and its corollary, war was immoral because it wasted energy and thereby lowered the level of civilization of those who engaged in it. Ostwald firmly supported this interpretation until his country entered World War I. Then he pragmatically altered it by stating that the highest civilizations made the most effective use of energy even when they were fighting wars. Wartime Germany, in Ostwald's judgment, was using energy more efficiently than her enemies in the Allied camp. Therefore, Germany continued to retain her superior civilized status while her foes were slipping toward barbarism.[13]

Ostwald's philosophy of science was far more sophisticated than these few anecdotes might indicate. He made a serious attempt to develop an energetic basis for all the sciences. In this attempt he opened his analysis within the confines of the physical sciences, shifted to the biological sciences, and concluded with a section on the social sciences and the role energy played in determining civilization.[14] Among the American intellectuals influenced by Ostwald's ideas was the historian and writer, Henry Adams. In a famous essay, "The Virgin and the Dynamo," Adams evaluated the effects of two great and different energies upon the course of Western civilization. The religious energy generated by medieval Catholicism in its veneration of the Virgin Mary had caused magnificent cathedrals to be erected and inspired poets, philosophers, and theologians in their creation of the culture of the Middle Ages. Opposing the energy of the Virgin stood the dynamo, a secular generator of energy, values, and civilization in the twentieth century. The force of electricity emanating from the steam-driven dynamo was every bit as mysterious and powerful as the religious force produced by the adoration of Mary. Yet, the cultures that sprang from these two generating sources were radically different. One was religious and unified; the other was secular and fragmented.

Henry Adams added further refinements to his commentary upon energy and civilization when he included the phase rule recently discovered by chemist Josiah Willard Gibbs. Adams wrenched the phase rule from its chemical context and applied it to the energy-civilization equation. In doing so he was convinced that he could reconstruct the entire curve of the history of civilization and identify precisely the phases or ages that constituted history. The first phase in human history, according to Adams, extended from the appearance of man on earth to the year 1600. During this initial phase religious energy determined the nature and direction of civilization. The coming of the steam engine heralded the end of the religious phase and the opening of the age of mechanical energy which was to last for three hundred years. The mechanical phase, in turn, was supplanted by the age of electricity that arrived in 1900 and would be extinguished within seventeen and one-half years.

The phases of human civilization were to become progressively and alarmingly shorter because the energy that drove them was limited and was dissipated more quickly. By 1917, said Adams, there would remain only one untapped source of energy—the ether that filled the entire universe. But ethereal civilization was doomed to perish within four short years and then the human mind would reach the ultimate limits of its possibility in the year 1921.[15]

The third twentieth-century figure in this account is the British chemist Frederick A. Soddy. Working with other prominent British scientists Soddy developed the disintegration theory of radioactivity, confirmed the transmutation of the elements, and was the first to advance the idea of the isotope. For this work he won the Nobel prize in chemistry in 1921—the ominous year that Henry Adams had chosen to mark the end of civilization. Not only was Soddy a pioneer in the study of the atom; he was also an early and enthusiastic proponent of atomic energy. If mankind was ever able to release the energy in the heart of the atom, wrote Soddy, he would have available to him a huge and inexhaustible supply of energy that would utterly transform society and lift civilization to heights hitherto undreamed of. An atomic-powered society of the future would reach such a peak of perfection that it could only be compared with the paradise we know as the

Garden of Eden. The gulf between the coal civilization of the 1920s and the atomic age of the future was to be equated, in Soddy's mind, with the vast difference that existed between the world of our cave-dwelling ancestors and the sophisticated Western societies flourishing in the early decades of the twentieth century.[16]

Soddy recognized two serpents, two sources of evil, bent upon destroying his atomic-powered Garden of Eden. The first was war. Suppose the energy of the atom was used for warfare and not for the advancement of civilization? Science-fiction writer H.G. Wells had already raised that possibility in a novel he wrote shortly before the beginning of World War I and dedicated to Soddy. In the Wellsian version of the future atomic energy was first released in bombs and only after the world was reduced to ruins and misery did the surviving humans build a new and glittering civilization using the energy of the atom for peace.[17] The other nemesis threatening future dwellers in Eden was the traditional economic system which hampered the equitable distribution of available goods and services. In attempting to solve this problem Soddy turned from chemistry to economics. His economics, however, was to be truly scientific for it was based upon the laws of thermodynamics. Wealth, Soddy claimed, was essentially the product of useful and available energy and the entire body of economic thought must be revised accordingly. Once the economic system was placed upon an energetic basis current inequities would vanish and a paradise would await mankind in the atomic future.[18]

During the years that Soddy was attacking the economic practices of his day and promising a new civilization based upon limitless energy supplies, two famous Americans—Henry Ford and Thomas Alva Edison—were dealing with questions of energy and economics in a similar fashion. At the time of the First World War the United States government had constructed a large hydroelectric plant for the production of fertilizer at Muscle Shoals, Alabama. The war ended before the ninety-million dollar plant could be used and in the 1920s there developed a national debate about the future of this expensive, but apparently useless, facility. Henry Ford offered to lease the plant, produce cheap electricity, and radically transform the economically depressed region. He planned to build a city there that would be 75 miles long and 15 miles wide. This new city might not quite rival Soddy's Garden of Eden but he promised that at least it would be larger than Detroit.

In December of 1921 Ford and Edison visited the Muscle Shoals site and were acclaimed by the populace as saviors of the region. There remained, however, the thorny matter of financing the undertaking. Ford proposed that money be printed by the government specifically for this worthy project. In an interview with the *New York Times*, Edison discussed the energy implications of Ford's financial scheme. He began the interview with the statement that gold might soon be abandoned as the basis for American currency. This claim was based on the erroneous notion that scientists had recently proved that lead was a compound and that gold might also be one. Apparently Edison was passing on a garbled account of Soddy's recent work on isotopes and the transmutation of the elements.

From gold Edison moved on to paper money where again he revealed his debt to Soddy. Edison observed that paper money was a sure sign of a highly civilized people who could accept bills and checks without the need to handle precious metal. Then he suggested that since energy was the true basis of money the federal government should print "energy dollars" with pictures of a hydroelectric plant prominently displayed upon them. These "energy dollars" would finance Ford's scheme for transforming bad farms and worn-out towns into a veritable utopia by means of the addition of abundant electrical energy.[19] A wise British prime minister once said that the surest way to get into an insane asylum was to study the money question. Soddy and Edison might well have heeded that advice, for neither distinguished himself in discussing possible reform of the monetary system.

Ford's plans for Muscle Shoals, never well developed, were killed when it became known that President Coolidge might have used Muscle Shoals as a card in a political game he was playing with the automobile manufacturer.[20] Nevertheless, the dreams of hydroelectric plants and a new society persisted to the time of Franklin Delano Roosevelt's administration when they became part of the Tennessee Valley Authority (TVA) program. These utopian dreams even spread beyond the boundaries of Tennessee and the United States. They traveled throughout the world as the TVA became the symbol of social and cultural transformation achieved by means of hydroelectric power. The TVA mystique was to reappear in such unlikely and remote areas of the world as Aswan, Egypt and the Mekong river delta in Vietnam. It would be foolish to argue that hydroelectric establishments have no influence on their social and intellectual milieu. However, it is also true that they have not created the revolutionary changes promised by their promoters. We have yet to create a new civilization merely because we harnessed the power of some wild river. There remains a considerable gap between utopian societies projected and the economic and ecological liabilities of power dam construction.[21]

The discussion of the scientific and technological origins of the energy-civilization equation must stop here. However, much more remains to be said on the subject of energy and civilization. One could pursue the theme in the work of Sigmund Freud, who linked sexual energy, and its sublimation, to civilization.[22] One might explore current anthropological and sociological thought based upon a link between energy use and cultural achievement[23] or review the astronomical speculation that ranks civilizations of yet to be discovered extraterrestrial beings according to their supposed access to different quantities of energy.[24] Finally, one might take notice of the fact that the introduction of new energy sources—coal, petroleum, the atom, the sun—have been invariably accompanied by highly exaggerated claims that they would be the basis for a new society and a higher civilization.

THE VALIDITY OF THE EQUATION

After we have explored the historical origins and diffusion of the energy-civilization equation there still remains the problem of its ultimate validity. Given the equation's technological and scientific background how valid is the basic assertion that energy consumption can be used as a determinant of civilization? One of the striking features of the equation is that slight increases or decreases in energy use cause large fluctuations in the level of civilization. If man uses less coal or electricity then he is surely doomed to wear animal furs, gnaw on bones, and pass his time shaping stone tools. On the other hand, if he only adopts solar, fusion, or some other new energy source then the gates of the Garden of Eden will be opened to him. We should be suspicious of a formulation that places mankind so precariously between apocalypse and utopia. And, we should be cautious in accepting an equation that does not reflect the fact that the vast increases in energy consumption over the past few decades have not necessarily enhanced our chances of reaching a new stage in civilization.

Perhaps one reason why our rapidly increasing energy consumption has not placed us upon a new plateau of civilization is the way in which we choose to use that excess energy. The crude formula linking civilization with energy has no place for questions of choice. It deals with energy expended per capita and does not ask if the energy was squandered on trivialities, wasted in destructive wars, or utilized to advance the social, moral, and cultural accomplishments we identify with civilization.

The failure to discriminate among the possible societal uses of energy can lead to absurd results. A modern British archeologist, Grahame Clark, proves my point. After stating that the low cultural status of savage societies arose from the small amount of energy at their disposal, he drew a disparaging comparison between savage and modern Europe. The total energy available in all of savage Europe, he announced, probably never equalled that of a single, modern four-engined bomber.[25] Passing over the obvious and embarrassing question as to how he had determined the energy available in savage Europe, we ask Clark: "Who is more civilized, the savage engaged in rock-throwing wars or the modern European who cunningly accumulates large amounts of energy so that he can build weapons for use against civilian populations?" Although it might be difficult to get a consensus on this question it is reasonable to assume that most would agree that it is not sufficient to calculate gross energy usage without asking specifically how, when, or where that energy is utilized.

Another weakness of the equation grows out of the vague way energy is defined within its context. There is no quarrel when the term is limited to the physical domain. But what are we to make of the analogical reasoning that led to the writing of serious essays upon moral energy, sexual energy, or religious energy? The culprits here are not only literary figures like Henry Adams but recognized contemporary scientists. For example, the author of a currently popular anthropology textbook routinely examines the various fuels and the kinds of societies they might help to establish. Having completed this examination, he casually announces that his next topic will be spiritual energy and the ways it is harnessed in prayer, sacrifice, ritual, and other religious practices.[26] Certainly neither religion, nor science, nor rational discourse is advanced by this blatant misuse of an important physical concept. The same process that helped to confer scientific respectability upon the energy-civilization equation also unfortunately laid it open to an indiscriminate, imprecise, and ill-informed use of the idea of energy.

If the energy-civilization formulation encouraged the loose definition of energy what did it do to the definition of civilization? Energy has its roots in the physical sciences so that no matter how it is misapplied the original concept maintains its integrity. Civilization, on the contrary, never has had the kind of precise determination we associate with an accepted scientific concept. Civilization has always been a value-laden word whose meaning has changed over time, being redefined again and again in order to meet current political, social, cultural needs or desires. It is an ill-fated formula that would attempt to link closely two such disparate entities as energy and civilization. Yet if one attempts to make the equation workable by focusing upon one nation at a given time and assuming that its people will agree on what is meant by civilization grave difficulties arise. When Stanley Jevons was predicting the imminent decline of British civilization in the 1860s he supposed that coal, iron, and railroads had raised England to the pinnacle of culture. Not so, responded the contemporary literary critic Matthew Arnold. Let us suppose, said Arnold, that 200 years from now England was to be swallowed up by the sea. Then when the rest of the world recalled England's greatness they would undoubtedly remember the age of Shakespeare as her golden hour and not the time of Alfred, Lord Tennyson and Queen Victoria. The Elizabethans managed quite well without the steam engine to produce a culture admired throughout the world.[27]

Should a twentieth-century opponent choose to debate Matthew Arnold over this matter he would probably draw upon statistics proving modern superiority in life expectancy, literacy, nutrition, public hygiene, speed of transportation, equality of opportunity, and so on. In short, he would shift the argument to the arenas of quality of life and economic growth. Matthew Arnold would reply to his modern critic as he did to Jevons by defining national greatness as that quality that excites "love, interest, and admiration for a nation and its deeds." And Arnold would be correct in doing so because throughout its history the energy-civilization equation has stressed the highest cultural acievements of man and not the more mundane aspects of his life. After all, the steam engine was praised not merely because it could pump drinking water to city dwellers but because it was a dispenser of culture.

A persistent critic would then respond that there just might be a connection between the availability of potable water and the creation of high culture. Must not the artist and scientist be fed, clothed, and sheltered before he can address himself to artistic and scientific affairs? There is no simple reply to this question. However, it calls for something far more profound than the energy-civilization equation which has often been put forth as the definitive answer.

There is a great danger in assuming that cultural attainments must wait upon the fulfillment of creature comforts, that man could not study the stars, think about gods, or ornament a piece of pottery until he had a full stomach, a roof over his head, and a wall around his city. The more historians and anthropologists learn about the early history of mankind the more they are convinced that science, religion, and art were always part of his existence and not refinements he cultivated after reaching a certain stage of economic surplus. Neither historical nor anthropological research supports the popular view that economic necessity is prior to, and prepares the way for, the moral, intellectual, and aesthetic life of man.[28]

In the final analysis it is not crucial that all of the criticisms brought against the energy-civilization equation in this essay be accepted. It is much more important that the equation be recognized as a pervasive, if often implicit, element in both popular and scholarly approaches to energy and society. If the energy-civilization equation is worthless and potentially dangerous it should be exposed and discarded because it supplies a supposedly scientific argument against efforts to adopt a style of living based upon lower levels of energy consumption. If it is a generalization of great truth and intellectual worth, then it deserves a more sophisticated and rigorous handling than it has received from its supporters to date.

NOTES

[1] Andrew Hardy, "Man's Age-Old Struggle for Power," *Natural History* 82 (October 1973): 82-86.
[2] John P. Holdren, "Too Much Energy Too Soon," *New York Times*, July 23, 1975.
[3] Aldous Huxley, "Progress," *Vanity Fair* 29 (January 1928): 69.
[4] Lynn White, *Medieval Technology and Social Change* (London: Oxford University Press, 1972): 79-134.
[5] John Farey, *A Treatise on the Steam Engine*, vol. 1 (London: Longmans, 1827), pp. 1-8.
[6] John Timbs, *Wonderful Inventions* (London: Routledge, 1868), pp. 186-187.
[7] Howard Mumford Jones, *The Age of Energy* (New York: Viking Press, 1970), pp. 142-144.
[8] Thomas Ewbank, *The World a Workshop* (New York: D. Appleton and Co., 1855), pp. 73-75.
[9] Stanley W. Jevons, *The Coal Question* (London: Macmillan, 1865).
[10] Thomas S. Kuhn, "Energy Conservation as an Example of Simultaneous Discovery," in *Critical Problems in the History of Science*, ed., Marshall Clagett (Madison: University of Wisconsin Press, 1959).
[11] Edward L. Youmans, *The Correlation and Conservation of Force* (New York: D. Appleton and Co., 1865), pp. xi-xlii.
[12] Edmund P. Hillpern, "Some Personal Qualities of Wilhelm Ostwald Recalled by a Former Assistant," in *Chymia*, vol. 2, ed. Tenney L. Davis (Philadelphia: University of Pennsylvania Press, 1949).

[13] Milic Capek, "Ostwald, Wilhelm," in *The Encyclopedia of Philosophy*, vol 6, ed. Paul Edwards (New York: Collier-Macmillan, 1967).
[14] Wilhelm Ostwald, *Natural Philosophy*, trans. Thomas Seltzer (New York: Henry Holt, 1910).
[15] William H. Jordy, *Henry Adams: Scientific Historian* (New Haven, Conn.: Yale University Press, 1952), pp. 121-157.
[16] Frederick Soddy, *Science and Life* (New York: E.P. Dutton, 1920), pp. 22-24.
[17] Herbert George Wells, *The World Set Free* (New York: E.P. Dutton, 1914).
[18] Frederick Soddy, *Wealth, Virtual Wealth and Debt* (London: George Allen and Unwin, 1926), pp. 49-68.
[19] *New York Times*, December 6, 1921, 6:1.
[20] Reynold M. Wik, *Henry Ford and Grass Roots America* (Ann Arbor: University of Michigan Press, 1972), pp. 106-123.
[21] James W. Carey and John J. Ouirk, "The Mythos of the Electronic Revolution, I and II," *The American Scholar* 39 (Spring, 1970): 219-241; (Summer, 1970): 395-424.
[22] Sigmund Freud, *Civilization and Its Discontents*, trans. Joan Riviere (London: Hogarth Press, 1930).
[23] Richard Newbold Adams, "Man, Energy, and Anthropology," *American Anthropologist* 80 (June 1978): 297-309.
[24] N.S. Kardashev, "Transmission of Information by Extraterrestrial Civilizations," *Soviet Astronomy* 8 (September-October 1964): 217-221.
[25] John Grahame Douglas Clark, *From Savagery to Civilization* (New York: H. Schuman, 1953).
[26] Philip K. Bock, *Modern Cultural Anthropology* (New York: Random House, 1974), pp. 275-277.
[27] Matthew Arnold, *Culture and Anarchy*, ed., J. Dover Wilson (London: Cambridge University Press, 1961), pp. 50-51.
[28] Adam Ferguson, *An Essay on the History of Civil Society, 1767*, ed., Duncan Forbes (Edinburgh: Edinburgh University Press, 1966), pp. xxi-xxii, and Claude Levi-Strauss, *The Savage Mind* (Chicago: University of Chicago Press, 1966).

5
The Human Factor
Eric Hoffer

From the early days of the Industrial Revolution intellectuals of every sort predicted that the machine would make man superfluous. Right now it would be difficult to find a social scientist who does not believe that automated machines and computers are eliminating man as a factor in the social equation. There is even an academic joke on the subject: Descartes said, "I think, therefore I am"; the computer says, "I think, therefore you are not."

The belief that the machine turns men into predictable robots is not based on experience or observation. It is an a priori assumption that blinds social scientists to what is happening under their noses. It prevents them from seeing that a triumphant technology is doing the opposite of what they predicted it would do. There is evidence on every hand that the human factor has never been so central as it is now in technologically advanced countries. It is the centrality of the human factor that makes industrialized societies so unpredictable and explosive.

In technologically advanced countries at present there is little that can be taken for granted. Children no longer automatically grow up to become adults nor do they automatically learn to read and write. It is no longer axiomatic that people are willing to work for a living or that they will work harder or longer hours for higher pay. There is no certainty that the end result of a course of action will be what it was reasonable to expect. So often it seems as if some cunning, spiteful power is playing tricks behind the scenes delighting in drawing from men's actions consequences least foreseen. The fear of militants exploding in the streets is preventing governments from making the right decisions both at home and abroad. The centrality of the human factor makes it impossible for free societies to have strong governments. Everywhere we look we see the paradoxes of the human condition play havoc with the best laid plans and the best intentions.

A further enhancement of the human factor came with the exhaustion of raw materials and sources of energy. In order to survive at present a country must not only know how to cope with social anarchy but also how to tap the creative capacity of its people. In the postindustrial age human, rather than natural, resources will be the wellspring of a country's wealth and vigor.

Social automatism is highest in societies dominated by iron necessity. In such societies the human condition is not allowed to inject its anarchic unpredictability into the social process. It is repressed and kept locked in the dark cellars of the individual's psyche. The hidden hand of necessity regulates and disciplines people, and the need for the deliberate management of men is minimal.

The nineteenth century, which was engaged in a Promethean effort to harness nature, gave little thought to the management of men. The ruling middle class could proceed on the principle that government is best when it governs least. Everyday life had a fabulous regularity. There was more robot-like behavior in the nineteenth than in the twentieth century. Millions of people went to work each morning and returned from work each evening with a regularity akin to the moon's tides. Obedience to authority was as automatic as a reflex movement. Social processes were almost as rational and predictable as the processes of nature which the scientists were then probing and elucidating. It was reasonable to believe in the possibility of a social science as exact as the natural sciences. Walter Bagehot wrote a book called *Physics and Politics*. There was also boundless hope, a belief in automatic ceaseless progress, which infused people with patience.

Then came the twentieth century. Have there ever been two successive centuries so different from each other as the nineteenth and the twentieth? The nineteenth century was stable, predictable, rational, hopeful, free, fairly peaceful, and lumpy with certitudes. Everything seemed to make sense: industrialization, railroads, steamships, exploration, emigration, empire building. America in particular was up to its neck in the purposeful action of taming a virgin continent. Even the bloodiest civil war made sense since it was fought to preserve one nation indivisible.

The twentieth century has been hectic, soaked with the blood of innocents, fearful of the future, crisscrossed with frontiers that prevent free movement, stripped of certitudes, unpredictable, and absurd. The history of the twentieth century is a succession of disastrous absurdities: World War I, the Russian revolution, the Versailles Peace Treaty, prohibition, the wild twenties, the Great Depression, the Roosevelt Administration (which tried to end the depression by killing pigs and dumping wheat), the Hitler revolution, World War II (which brought the Russians to the bank of the Elbe), the Holocaust, the absurd 1960s, and now the Carter presidency.

What was it that made the twentieth century so different from the nineteenth? World War I was a sharp dividing line between the two centuries. But it was not the savagery of the war and the frightful waste of lives that made the twentieth century bloodsoaked and absurd. After all, the prelude to the rational, peaceful nineteenth century was the 20-year hemorrhage of the revolutionary and Napoleonic wars. No, it was not World War I itself but its aftermath which shaped our terrible century. Without the breakdown of czarist Russia and the humiliation of Germany by the Versailles Peace Treaty there would have been neither a Lenin nor a Hitler revolution. The war also saw a wholesale slaughter of certitudes and the loss of faith in progress.

There have been many attempts to show that World War I was an unavoidable product of the nineteenth century. Psychiatrists in particular delight in enlarging on the spiritual breakdown which preceded the war; that the war only gave the *coup de grace* to a spent cultural and political structure. We are told how Einsteinian physics, cubist painting, and other innovative tendencies introduced a pernicious relativism which made Europe ripe for an apocalyptic war.

Actually, we have the testimony of highly reliable observers on the fabulous stability and hopefulness of the prewar decades. To Alfred North Whitehead who was immersed in the new physics the period 1880-1914 was one of the happiest times in the history of mankind. He often thought what a wonderful world to live in his children would have. The generation that blundered into World War I had all the certitudes and illusions of the nineteenth century. They took civilization, progress, monetary stability, and rationality for granted. It would have been difficult to find in 1914 anyone who felt that he was living in the last year of a dying age.

It is impressive how logic and hope kept nineteenth-century thinkers from contemplating an unpleasant, let alone apocalyptic, denouement as the fulfillment of the Industrial Revolution. Few in the nineteenth century were aware of the explosive irrationality of the human condition. No one suspected that once nature had been mastered and the hidden hand of scarcity no longer regulated and disciplined people, industrial societies would enter a psychological age in which man would make history and also become a threat to mankind's survival.

In the rational and hopeful climate of the nineteenth century there was no place for forebodings that the paradoxes of the human condition would combine to turn the successes of an industrial society into failures. No one foresaw the disintegration of values and the weakening of social discipline caused by the elimination of scarcity. A logician like Marx could not foresee the downfall of capitalism by ever-increasing affluence rather than by ever-increasing misery. Hardly anyone in the nineteenth century foresaw the chronic unemployment and the loss of a sense of usefulness caused by increased efficiency. No one feared that rapid change would upset traditions, customs, routines, and other arrangements which make everyday life predictable. Finally, no one foresaw that the education explosion made possible by advanced technology would swamp societies with hordes of educated nobodies who want to be somebodies and end up being mischief-making busybodies.

Strangely enough, the one thing the nineteenth century viewed with alarm was the masses. Many saw the masses as the womb of chaos. Jacob Burckhardt thought the masses loathed stability and continuity. They wanted something to happen all the time and their clamor for change would topple all that is noble and precious. Walter Bagehot believed that once the masses were given political power only education and prosperity could preserve social stability. To Freud it seemed that the individuals composing the masses support one another in giving free rein to their indiscipline. No one had an inkling that anarchy when it came would originate not in the masses but in violent minorities including the minority of the educated. Everything that was said about the anarchic propensities of the masses fits perfectly the behavior of students, professors, writers, artists and their hangers-on during the 1900s. The masses are the protagonists of stability, continuity, and of law and order.

It is curious that Disraeli should have had a truer view of the nature of the masses than his liberal contemporaries. He sensed the conservatism and patriotism of common people. Could it be that being a genuine conservative Disraeli was more attuned to the eternal verities of man's existence? Considering also how timely and relevant Disraeli's ideas about what makes nations strong and great it is legitimate to wonder whether you have to be a conservative if you want to be up-to-date tomorrow.

The twentieth century saw not only the fulfillment of the Industrial Revolution but also the fulfillment of words planted in the preceding century. There is hardly an atrocity perpetrated in the twentieth century that was not advocated by some intellectual in the nineteenth.

The nineteenth century was dominated by men of action. The intellectuals just talked, and no one expected their savage words to have consequences. The intellectuals entered the nineteenth century convinced that it was going to be their century. Have they not made the French Revolution? They saw themselves as the coming ruling class. But things turned out differently. The Industrial Revolution gave power to the lowbrow middle class and the intellectuals were left out in the cold. The nineteenth century reverberated with the genocidal ravings of intellectuals against a mean-spirited, usurping bourgeoisie.

Now, it is the predicament of the middle class that although it excels in mastering things, it is awkward and almost helpless when it comes to managing men. It is the only ruling class that does not dabble in magic and has no rituals. Thus when, with the consummation of the Industrial Revolution the human factor became more and more central, the middle class, drained of confidence by World War I and the Great Depression, found itself in deep trouble. The stage was set for the entrance of the intellectuals.

To an intellectual power means power over men. He cannot savor power by moving mountains and telling rivers whether to flow. He is in his element commanding, coercing, brainwashing, and in general making people love what they hate. He glories in the role of medicine man and charismatic leader and feels God-like when he makes words become flesh. Thus he has made the twentieth century a century of words par excellence. In no other century have words been so dangerous. A failure to recognize this fact can have disastrous consequences.

The Occident paid dearly for refusing to take Lenin's and Hitler's words seriously.

Though we find it hard to believe that "In the beginning was the word" and that words created the world, we have seen words ignite genocidal passion and we are ready to believe that words may destroy our world.

Viewed from any vantage point the nineteenth century was a sharp historical deviation. About 150 years ago the Occident was catapulted into a trajectory away from the ancient highway of history. We can now see that the trajectory is a loop that turns upon itself and is curving back to where it started. We can see all around us the lineaments of a preindustrial pattern emerging in postindustrial societies. We are not plunging ahead into the future but falling back into the past. The explosion of the young, the dominance of the intellectuals, the savaging of our cities, the revulsion from work are characteristics of the decades which preceded the Industrial Revolution. We are returning to the rutted highway of history and are rejoining the ancient caravan.

The significant point is that the people who are rejoining the ancient caravan are not what they were in preindustrial days. They are more dangerous. The unspeakable atrocities of the twentieth century have demonstrated that man is the origin of a great evil that threatens the survival of mankind. The central problem of the postindustrial age is how to cope with this human evil.

It is conceivable that if the exhaustion of raw materials and sources of energy make it imperative for a society to tap the creative energies of its people it may, in doing so, also tap a new source of social discipline. For the creative individual, no matter how highly endowed, must be hardworking and disciplined if he is to accomplish much. There is no invention that will take the hard work out of creating. Moreover, since the creative flow is never abundant, the creative society is likely to be disciplined by chronic scarcity.

The trouble is that the coming of the creative society will be slow and faltering, and we must find other defenses against evil.

As things are now, it may well be that the survival of the species will depend on the capacity to foster a boundless capacity for compassion.

In the alchemy of man's soul almost all noble attributes—courage, love, hope, faith, duty, loyalty—can be transmuted into ruthlessness. Compassion alone stands apart from the continuous traffic between good and evil proceeding within us. Compassion is the antitoxin of the soul; where there is compassion even the most poisonous impulses remain relatively harmless.

And compassion is uniquely human. Nature has no compassion. It is, in the words of William Blake, "a creation that groans, living on death, where fish and bird and beast and tree and metal and stone live by devouring."

Compassion seems to have its roots in the family. We think of those we love as easily bruised, and our love is shot through with imaginings of the hurts lying in wait for them. Parents overflow with compassion as they watch their children go out into a strange, cold world. Now it is conceivable that the present weakening of the family may allow compassion to leak out into wider circles. So, too, the cultivation of *esprit de corps*, which is the creation of family ties between strangers, may aid the spread of compassion. But the question is: Can we make people compassionate by education?

It is natural to assume that the well educated are more humane and compassionate than the uneducated. But the reverse seems to be true. When Ghandi was asked what it was that worried him most, he replied, "The hardness of heart of the educated." We have seen the highly educated German nation give its allegiance to the most murderously vengeful government in history. The bloody-minded professors in the Kremlin, as Churchill called them, liquidated 60 million men, women and children. We have also seen a band of graduates of the Sorbonne slaughter and starve millions of innocents in Cambodia and Vietnam. The murder weapons that may destroy our society are being forged in the word factories of our foremost universities. In many countries universities have become the chief recruiting ground of mindless terrorists.

Now, I have never been a teacher or a parent, and my heart is savage by nature. I am therefore unfit to tell people how to implant compassion. Still, I have the feeling that, perhaps, the adoption of a certain view of life might be fruitful of benevolence and compassion.

We feel close to each other when we see ourselves as strangers and outsiders on this planet or when we see our planet as a tiny island of life in an immensity of nothingness. We also draw together when we become aware that night must close in on all living things, that we are condemned to death at birth, and that life is a bus ride to the place of execution. All our squabbling and vying is about seats in the bus, and the ride is over before we know it.

II

Adapting the Institutional Frame of Technology

6
Two Kinds of Light From Science
Philip Morrison

Perhaps more important than the light of the wonderful electric lamp of Edison is the second kind of light offered us by science. It is what we call insight, sometimes even illumination: the understanding of ourselves, our origins, our place in the universe, an understanding that science slowly builds. Fifty years ago, at a time after Einstein but before fission, after the chromosomes but before DNA, after the rise of radio but before electronic computers, the philosopher Whitehead put it eloquently:

> This quiet growth of science has practically recolored our mentality. . . . The new mentality is more important even than the new science and the new technology. It has altered the metaphysical presuppositions and the imaginative contents of our minds; so that now the old stimuli provide a new response. . . . This new tinge to modern minds is a vehement and passionate interest in the relation of general principles to irreducible and stubborn facts. All the world over and at all times there have been practical men, absorbed in "irreducible and stubborn facts"; all the world over and at all times there have been men of philosophic temperament who have been absorbed in the weaving of general principles. It is this union of passionate interest in the detailed facts with equal devotion to abstract generalization which forms the novelty in our present society . . . the tradition which infects cultivated thought. . . . It is the salt which keeps life sweet.[1]

OUR HUMAN PLACE IN SPACE AND TIME

As old as myth, as old as imagination itself, is our need to learn our part in the great theater of space and time. When Dante wrote, not a thousand years back, he knew that this earth was a great globe. But he thought that upon it one could find the physical gates to Purgatory and to Hell, somewhere far from Rome and Jerusalem. The modern world came into full being, it may be, when the scholars at last understood that our limited world held among its volcanos no such terrible grandeur, that it was a rocky ball flung in space, one planet among the planets, no earthier, no less shining, than starry Jupiter or red Mars. The earth is not at the center of all things, not cosmical hub, but our own limited and physical home. We came to see the sun as a unique luminary only for us, but really one star among a myriad stars, two hundred million quite like it in our Milky Way.

Our place in space, almost all that the last lines summarized, was already well known to Isaac Newton. But he remained lost in time. He held that our world was about six thousand years old, back to Eden. I myself have walked in human habitations three times older than that, in the wonderful painted caves of France and Spain. We know now that the earth is ancient, almost five billion years old, although the fixtures of our maps, the Rockies, even the Atlantic Ocean, are much younger, hardly pushing a couple of hundred million years. Life is at least three billion years old on earth, and for all but the last one-sixth of that time it remained solely marine, solely the creature of water. All the fossils of the museums which speak of an ancient life on land are novelties, come on earth in its maturity. If the history of earth were mapped in a time map and the scale chosen was one year for all the duration of earth, land life would appear about Thanksgiving, the Atlantic Ocean somewhat before Christmas, the Mediterranean would fill with sea water about noon of New Year's Eve, and our first ape-like ancestors would leave the trees as dusk fell that last day. Father Abraham left Ur half a minute back. The telescope is about three seconds old, and Edison's lamp under a second old out of the full year. Such a lengthy prelude to our own brief appearance cannot fail, it seems to me, to condition our deepest instinctive ideas about our shortcomings and our hopes; to be sure, it cannot fix them explicitly. We are plainly newcomers, children still, in the procession of life and being.

We know more. We are tightly woven into the web of life. All human beings are quite literally cousins; a simple estimate suggests that on the average we are fifteen-or-twentieth cousins, sharing grandparents that far back, true blood kin. More than that, we are blood kin similarly to the apes—no surprise—to the fishes, even to an oak tree! Life is a single family, whose pedigree is written in molecules.

THE THEMES OF HUMAN HISTORY

So much is the fruit of the natural sciences. Science has learned a good deal, though less certainly, about human existence itself. All that I offer in this line is probable, though not so sure as what stands above.

Our human family has a history of community, as well as a physical kinship. Not so long ago all our forebears were hunters and gatherers, before Adam ever delved or Eve spun. Most of human history, if we could write it fairly, was spent in such a life, a life of skillful little bands roaming the landscape, without rulers, without cities, without money, without war, without the sowing of crops or the breeding of animals, save perhaps the dog. We were more alike in occupation and in appearance then; out of Africa we wandered, in the end to the steppe, the edge of the ice, the rain forests, and slowly we turned a little different in stature and feature, some becoming paler, some darker, from our brown African Rift forebears.

Human beings can thrive only in society, but we are by no means the only organisms to so evolve. Consider the termites, an ancient form closely related to the roaches. They are blind farmers, who in the dark sow, tend, reap, propagate (outside) and finally depend upon their particular domesticated staple protein food, a species of fungus found nowhere else save in termite gardens. They are talented, if conservative, architects, with a sure mastery of arched vaults and of successful air conditioning. They follow specialized occupations, including gardening, nursing, soldiery, palace attendance, and common labor. It is a plausible conjecture that their cities, with populations of hundreds of thousands at a time, lasting over generations of inhabitants, are limited in civilized accomplishments, not by any intrinsic limits of such blind evolution, but by its painfully slow rate. For culture which must grow almost by natural selection alone, without transmitting learning, language, or many artifacts, is very slow to change. The universe is not old enough for termite society to have elaborated as has our own. Termites live without individual rights. A slightly injured termite is solicitously groomed by his neighbors, who worry and stroke the incipient wound. If by chance it grows a little worse and droplets exude, it is quite likely that the concerned neighbors will at once tear their ailing comrade apart, to eat him with dispatch. Termites are usually quite short of protein. Their dead are not buried.

THE HUMAN MIND AND OUR SOCIETY

The termite individual is profoundly different from the human individual. Termites behave according to a subtle inbuilt code of behavior, locked deep in the genes. They learn little, with only tiny brains. Humans learn much; whatever the potentials established genetically, their individual life experiences give them a great deal from language upward. No two persons are as alike inwardly as are all termites. The rights of the human individual can be said to rest upon the uniqueness of every person; learned and innate behavior add something unique to the whole. On the other hand, our long infancy and the paramount importance of language for the mind, to say nothing of all the richness we take in from every human culture, imply that human beings are dependent upon society. No abandoned infant is likely to grow to a mature citizen, even if the wolves take a hand. Human society lies between the termites' pure collectivity and the romantic individualism of our hero legends; human beings are individuals, but with the rights and responsibilities of citizens bound into a polity. All durable groupings of men, women, and children sooner or later strike some intermediate balance between social structure and individual claim, a balance more subtle and shifting than any pattern we can see in nature.

It seems probable that the slow polishing of the millenia gradually brought the society of the hunters and gatherers to a long-lasting and stable form—past what crises and sudden mutations we do not know. Our industrial-agricultural world has had no time for that adjustment yet; we have real grounds for hope in our immaturity, along with a growing need for progress to that end.

We have no inborn occupational adaptions as visible as those of termite castes, but always we have had a division of society. That division has increased over the long record of human social evolution, but it was there long ago in the hunters' world. They appear to have had shamans who were at once performers and therapists, and other artists, as well, perhaps even astronomers who transmitted the lore of the sky. It has always been clear that the increased complexity of human society has in part reflected the increased differentiation of human tasks—differing experiences, aptitudes, early training, social constraints—among human beings. Once more we see a deep-seated tension in our social nature between the uniqueness of each mind and the need for social wholeness. Our times more than any others are taut with that pull.

MORE LIGHT

These statements embody no equations, no intricate logic. They are the simplest and perhaps the strongest of the results of that way of knowing which is called science. I believe they are indispensable—indeed, there is much more quite as interesting and important which I have not mustered here—as the tools of thought for everyone who dwells in this our world. They provide, not a single road to a self-image of our place in the great scheme, but a set of landmarks among which many persons and many groups can seek and find many paths.

They are perhaps the basis of a structure of myth, and so they can be held. Myth acquires strength in many ways, but not by an open examination of all that is known, general principle and stubborn fact. It leans more on the authority of seer or sayer. Science cannot long remain closer to authority than to fact; that would belie its nature. In our world, science, and the technology which lives with it, must do more than tell the tales. For it is not belief alone which makes science, it is always shared understanding, tentative and provisional until conviction follows the long working out.

There is no doubt that more people today think of pushbuttons and products than of circuits or substances. It has been much easier to distribute the product of high technology than to share the understanding of its roots. Edison's light follows the wiring crews; but insight does not go so simply with the load.

No more dangerous gap in the strained fabric of the modern world can be found than the gap between those who knowing bear the powers of technology and those who hear only the distant sounds of myth. It is the plainest of pitfalls that there are more who watch a television screen than have been brought to grasp the idea of a rate of change, more who drive autos that grasp the meaning of energy and its degradation. We can survive this social gap over the long run only by chance; if we are to survive it by wisdom, we need to make sure that the networks we build can power both kinds of light. The provision of insight is more and more becoming a charge upon the distributors of the goods of technology. For without it, the use of technology in war and peace can bring only division, not that mutual trust and wide cooperation which are a prerequisite for social survival. It is late, but not too late.

Back again to Whitehead, fifty years ago:

> Modern science has imposed on humanity the necessity for wandering. Its progressive thought and progressive technology make the transition through time from generation to generation a true migration into uncharted seas of adventure. . . . It is the business of the future to be dangerous; and it is among the merits of science that it equips the future for its duties. The prosperous middle classes who ruled the nineteenth century placed an excessive value upon placidity of existence. They refused to face the necessities for social reform imposed by the new industrial system, and they are now refusing to face the necessities for intellectual reform imposed by the new knowledge.[2]

The danger of the future is clear enough. What is yet unclear is how we can bring science, both its strongest results and its best means, into every reflective mind as we have carried its fruits to every home fed by a transformer. Until we make progress to that end, we have not adequately equipped the future.

NOTES

[1] Alfred North Whitehead, *Science and the Modern World*, The Lowell Lectures, 1925 (NewYork: Macmillan, paperback ed., 1967).
[2] Ibid.

7
Technology and Socioeconomic Innovation
Simon Ramo

Is our scientific and technological knowhow being applied adequately where there is strong promise of high economic and social rewards? To this important question the answer is "no." We are becoming slower, more timid, and less innovative in applying science. While our national capabilities in science and technology are unsurpassed, we are employing them less and in a timid, partial manner. When I refer to putting technology to work I do not of course mean unthinking applications or misuse of technology on projects the public does not in the end really want. Also, even perfection in the selection and implementation of technological programs would not guarantee a healthy economy and a happy society. But without a strong technological foundation, society's needs will remain unsatisfied. Science and technology are necessary if not sufficient for national welfare.

World economic health is fostered by each nation's supplying what best it can to other nations and trading with them in return for their most suitable items. Continuing advances in science and technology are fundamental to our ability as a nation to engage in this manifold process of exchange with the rest of the world. High technology exports such as aircraft, computers, petroleum equipment, and chemicals are presently the mainstays of the positive side of the United States trade balance. The other strong contribution of this kind comes from agriculture, mainly because our fertilizer, farm machinery, pesticides, and other food technologies lead the world. To maintain our trade position will require continued technological advance. We used to be ahead of others in expenditures for R & D as a fraction of GNP. We are no longer, despite the potential for productivity increase through application of new technology. Unfortunately, the rate of productivity increase for the United States is now lower than that of every other major industrial country.

INNOVATION AS A SOCIAL-ECONOMIC-POLITICAL PROBLEM

Why this deceleration in innovation? One reason is the recent wave of "antitechnology" public opinion in the United States. A substantial part of our citizens equate technology with the devil. We have been crowded into congested cities. Television delivers vapid, violence-laden programs. The automobile is an engine of death that fouls the air and snarls us in traffic. We waste money landing a man on the moon when our airports are less and less safe. The atom bomb may destroy civilization and the nuclear reactor may poison our environment. To recite such a litany is, of course, to list only the "bad" while overlooking the "good." These tools of man are made valuable by his proper use and threatening by his misuse. A broad, indiscriminate bias against technology would impair our overall ability to reach objective, sound, nonemotional decisions in every domain where it is involved. The majority of the public in the United States does not seem to understand the necessity for a strong program of advance in technology to meet its own insistence upon a growing standard of living, low unemployment, and a tolerable rate of inflation.

An even more serious limitation upon the use of science and technology is public confusion over the role of private enterprise and government. Many people are convinced that "big business" is socially irresponsible. Whenever a problem surfaces they are certain that business will try to exploit it, by seeking unconscionable monopoly profits at the expense of consumers. They look to government alone, not at all to the private sector, to provide solutions. Since technological advance is most often identified with big business, they reason that such advance yields unneeded products and higher prices.

Another segment of the voters is fed up with government spending and big government generally, which they see as a huge and increasingly incompetent and inefficient bureaucracy. These citizens are likely to regard government-sponsored scientific research as wasting dollars by the billions and major technological programs as boondoggles duplicating what should be left to private enterprise. As a capping indication of the degree of confusion, many, it would seem, hold both these extreme views at once. They distrust both the private sector's and the government's involvement with science and technology. This climate of "plague on both your houses" is damaging because unless we have a high degree of cooperation between the government and the private sector, and public understanding of the need for such cooperation, continued technological advance in the United States will become next to impossible. Because the government is so important in maintaining the environment for technological innovation, and because our nation's capacity for innovation resides almost entirely in private industry, the advance of technology depends on keeping these sectors together in a harmonious ensemble: government policy, private investment and performance by corporations, and public support.

In the United States today we are afflicted with a severe mismatch between the high potential of technological establishments and the lower potential of our social-political progress. The problem is not one of science or technology in isolation. We are simply not organized to use our tools to the fullest. It is the interface between technological and nontechnological factors that is controlling, and critical. We are confronting an unsolved "systems" problem. In choosing where and how to apply science and technology it would be helpful to have clear national goals. When we find it hard to articulate what kind of society we want, it is naturally hard to pinpoint the most effective ways to use science and technology. Satisfactory decisions stem from an awareness of trade-offs and options. The "good" that can come from specific advances needs to be compared with the "bad."

We do not rely exclusively on either government or business in the United States. We have a hybrid economic structure and seek to balance private and government action by assigning them different tasks. The free enterprise system of private capital at risk seeking a return has served well for two centuries to connect the needs and desires of citizens with many of the capabilities of technology, releasing an immense stream of products and services. Why does not the free market automatically solve the system and organizational bottlenecks just referred to? Why is it not the complete answer to the question of how best to put science and technology to work?

Part of the answer is that inflation, government spending, tax policies, and overregulation have impaired earnings and lowered the funds available in industry for plowback to create technological advances. Furthermore, the increasing likelihood of inadequate return militates against risking limited discretionary funds to exploit technological opportunities. After the small difference between selling price and costs is used to pay taxes, the interest on borrowed money, dividends to the shareholders, and the inflated costs of replacing depreciating facilities, too little is left to improve methods and develop new products. So dislocating have been the effects of inflation on cash flow that many corporations have overborrowed. They now find themselves with too high a debt in relation to their equity capital and net earnings. Since this condition is paralleled by high interest rates, their debt expense is extraordinary. Many companies' stocks are selling on the market at less than the book values of their assets even though inflation would prevent replacing those at their started book values. This means that new equity capital is as hard to come by as are earnings for reinvestment. This explains why few new corporations are being formed now to develop positions in new technological products and why mature companies are tending toward safer, short-term payoff, incremental advances.

Unfortunately, it is also true that the most needed and potentially beneficial technological projects increasingly call for investments well beyond the net worth of individual corporations, and they are not allowed to team up to share risks. For projects of this kind the "risk-to-return" ratio is too great, the losses on starting up are too high, the time to turn around to an eventual profit phase is too great, and the dependence of success on political decisions is forbidding, including international cooperation, which is less constructive than idealists are likely to suppose.

The important drag on innovative effort in United States industry is certainly not a lack of creative, inventive talent, but rather insufficient incentive to invest funds to back it, largely because inflation, high taxes, high costs in dealing with government regulation, high consumption-to-investment ratios, and the high ratio of public to private sector spending are just too discouraging. Given the state of the economy, no government program to stimulate more innovative effort and R & D in private industry could be nearly so effective as a commitment to lowering inflation and to free discretionary funds by lowering taxes. If government could create changes such as these, the seeds of science and technology would readily yield copious crops.

In considering the foregoing assertions, it is well to recognize one limitation on our national capability that the efforts of private industry can help to lift. This is the shortage of scientists and engineers experienced in certain critical fields. At all times we experience talent supply problems in various areas of science and technology. At one point, for example, the nation's goals called for more experts in space technology than could be found through the labor market. By imaginative conversions from other fields of expertise, accelerated educational programs in universities, and the enlistment of mature technologists, the shortage was overcome. A healthy, profitable, and growing private sector, together with sensibly chosen government projects, can act to meet requirements for human resources.

REALIZING THE PROMISE OF TECHNOLOGY

Let us now consider a number of examples that illustrate the dependence of technological advance on economic and political factors. To consider a specific example of the interplay of technology and economic-political considerations, if existing oil wells in the continental United States were equipped with submersible pumps for secondary and tertiary recovery wherever warranted, our oil reserves would be greatly increased, even doubled by some responsible estimates. Such installations are thoroughly practical from a technical standpoint. If the price per barrel were higher than present controls permit for "old oil," new investments would readily bring us tens of billions of dollars worth of additional petroleum. The critical pricing parameter, to make the additional investment sound, lies outside the free market. It is a political matter. Pricing involves a contest among numerous influential constituents with wide differences of opinion on the issue. The private sector will not produce the required funds without a change of public policy.

Improved Urban Transportation

Almost every large city in the United States could gain by installing a first-class system of public transportation. The technology would have to be carefully matched to each city's layout, industry pattern, population and employment configuration, and to such general qualities as its economy, social conditions, and life-style. The economic gains, assuming sound applications of the correct technology, could be prodigious. Perhaps half of our population could save hours each week over present means for getting to and from work, traveling at less expense, with less energy, fewer accidents, and less air pollution.

Is it remotely feasible for an American technological corporation to invest its resources in hopes of developing, selling, and finally earning a return from mass transit systems in our cities? Consider Los Angeles. Imagine that some private company a few years ago decided to devote several years of effort to fitting technology to the social and economic needs of this city and designing a multi-billion-dollar transit system. It would, before progressing further, have to work out all practical aspects of such subsystems as the vehicles, controls, communications, energy distribution, safety, terminals, maintenance, and environmental features. For all of this an expenditure of one hundred million dollars would be only a beginning. In order to bring prices down it would have to develop positions on many of these variables in other cities. It would then have to design, build, and tool up several plants to start producing the required hardware, refine it by full-scale testing, and manufacture parts in some quantity to be able to be sure of price and quality before taking on contracts for shipment and installation. If the company had gone this far, with an investment of more than one billion dollars, would it have succeeded in selling its system to Los Angeles? There is not even such an entity, organized as a single market for government procurement purposes. The "risk-to-return" ratio would be absurdly high for the private corporation we imagined electing to enter this field. Even assuming eventual success, the time to payoff would be too long and the approach of units of government to essential questions of pricing would be too unpredictable.

ENVIRONMENTAL TECHNOLOGY

Done suitably, with a mixture of creativity and common sense, the depolluting of major lakes, rivers, harbors, and coastal waters stands to yield a good return on investment in terms of health, quality of life, and economic gain through preservation of natural resources. To achieve practical results, much more scientific and technological work is needed to understand pollution phenomena more deeply, develop superior nonpolluting approaches to use of the waters, and produce specialized equipment. Learning quantitatively and qualitatively about industrial effluents, power utility outputs, transportation pollutants, and city sewage would have to be suceeded by an effort to understand the costs of minimizing, limiting, and altering these environmental impairments. The decision makers—in the end the public—must be able to see the options and compare the benefits with the price to be paid.

What company, or group of companies, would invest in a private overall systems solution to depolluting Lake Erie? This would mean setting out, with its own funds, to appraise a vast range of interactions and to develop new chemical processes, superior purification and fuel burning methods, optimum energy use techniques, and the rest. Many specific and difficult science and engineering advances would be vital for an intelligent, broad changeover to an integrated lower pollution system. But in what context should the individual technological efforts be chosen? What criteria should be used? Even if a combine could be created of, say, five large corporations that among them possessed all the technical expertise required, how could they presume to design and install a system that would affect the economy and social makeup of the industrialized communities of millions of citizens distributed around the lake?

If some private combine were to invest billions to develop a sound solution, it is not clear who its customer would be. To be sure, the companies taking the risk might offer a proposal to install superior waste treatment equipment to a city contributing pollution to the lake, but why should that city pay to pollute less unless all other polluters agree to do their part in a balanced program? The polluters would not only have to agree; all would have to possess the funds to buy their pieces of the system. Finally, suppose the numerous independent segments of the "customer" come together by some magic. That is, they decide to buy and stand ready with the funds to contract with the entrepreneurial syndicate to install the entire developed system. Would not the Justice Department be expected to stop this combine of private corporations, on grounds that their combination enabled them to dominate the market? Even if it escaped this threat, a second such team would not be likely to want to enter as a competitor.

Electronic Brainpower

Electronic information systems can now absorb, store, categorize, process, move, and present information in vastly higher quantities, with greater speed, yet with radically reduced cost and more reliability and accuracy than attainable ever before. This "synthetic electronic brainpower" can make human beings smarter at their jobs. We all spend much of our time doing something with information.

By a new man-technology partnership in information handling, the potential for increasing the value of hours of people's time, and hence the productivity of the nation is tremendous. Where workers' output could be improved to the extent of several hundred dollars each year by the application of this technology it would justify the investment of at least one thousand dollars per worker for the purchase of hardware and software. Such a mass application of electronic information technology would add up to hundreds of billions of dollars.

As an example of a market application that could induce change of this magnitude, consider electronic funds transfer, the use of electronic signals to replace the present production and flow of billions of pieces of paper a year for control and accounting in money transactions. Bills could be paid, orders placed, deposits to accounts made, interest charged or credited, taxes extracted, and funds assigned to or moved from individuals, companies, and governments to one another—all in accord with electronic signals. But the full realization of this lowered paperwork society would require a revolution in the rules governing the flow of money. The government would have to redefine banking, decide who could be allowed to profit from holding and moving money, standardize interest rates on funds stored, create privacy and security regulations, and set standards for signal transmission and equipment. In order to achieve electronic funds transfer we must first solve a challenging organizational problem, one requiring government and industry to team with science and engineering. It is not technology, but the rate of progress in our organizational skills, that determines how rapidly society can obtain the benefits in prospect.

Space Technology

The United States is ahead of other nations in most aspects of space technology. In coming decades, economic cost-effectiveness improvements can result from implementing large-scale satellite systems for telephonic communication, television broadcasting, airline navigation, computer-to-computer communication of business data, educational broadcasting to schools and homes, and a host of other service arrangements. The rate at which these developments are introduced will depend on how government-industry arrangements are handled.

Consider specifically a system to plot and examine the earth's resources from satellites, working with a ground-based network of communication equipment, computers, and data analyzers. A private entrepreneur, in the form of one or more private companies, would not create and operate a system, relying to any extent on government-controlled facilities and equipment, without a government-sanctioned position, which would expose it to regulation. The price of the information made available would have to be controlled and the whole could never be a "private enterprise" project.

To take another example, consider the use of satellites in space as precision "artificial stars" to provide greatly improved airplane navigation. By using appropriate satellites, each on highly accurately predictable trajectories, economical and low-weight electronic equipment in airplanes could track airplanes' locations with a degree of accuracy not possible today. To bring the new system into being would be an organizational problem involving airlines, airports, the regulatory agencies, the military, pilots' unions, and many more, including counterpart groups in foreign countries. The bottleneck in this useful prospective development is not technology but organization, and this has been true for years in some fields.

Energy

Controlled thermonuclear fusion has the possibility of providing us with cheap and plentiful energy. Many millions of dollars have already been spent trying to understand the underlying physics. It is estimated by most experts that decades would be required before all the remaining physical science and engineering details could be worked out so that this approach to energy would be ready for application. It might never be feasible. The total costs in the start-up phase are on the order of billions, perhaps tens of billions, of dollars. Amounts of this magnitude, time durations in decades, and doubts about feasibility carry us beyond the boundaries of areas suitable for speculative investment by private entities.

By all-out use of science and technology and the full cooperation of government and industry we could have a plentiful supply of energy for centuries by using our huge reserves of coal. Existing technological beginnings can be advanced in order to desulphurize coal, produce liquid and gaseous fuel from it, mine it more safely, and even use it in situ to generate electricity without mining it. The resulting energy supply would be higher priced than today's oil, in part to write off the costs of technological advances, but the funds to pay for the fuel would stay in the United States.

Success requires both the knowhow of private industry and enormous amounts of capital placed at risk. A corporation choosing to make a major entry into synthetic gas or liquid fuel from coal by new technology would have to commit funds in the range of billions of dollars and allow decades to pass before a return on the investment could be realized. The complete system needed for much greater utilization of coal in the United States involves a host of private and public organizations that are semi-autonomous. It is not difficult to understand why we are not moving rapidly in applying new coal technology.

Few doubt that energy will be more costly in the era ahead. But with a plentiful supply at the higher price level—in contrast to a subsidized, low-price, and low-quantity supply—the United States would prosper, mostly because the gains in economic growth and stability would outweigh the penalty of assigning a higher part of GNP to energy costs. A national program could be designed around the realistic expectations that the price of energy would increase because additional private investment would become available and private knowhow would be used to guarantee the supply, with the return on investment justifying the program. We would have to abandon the politically popular but unrealistic and inconsistent practice of susidizing low-priced energy.

Food and Nutrition

Nearly every nation in the world faces food insufficiency. The United States, with its resources and knowhow, has a special opportunity here. We possess an unmatched combination of land, climate, water, mechanical equipment, and chemical knowledge. Science and engineering advances could improve every aspect of food and nutrition. From food to mouth, we have strong positions in all pertinent technical areas. When, however, we consider the interactive network of land use, energy and water supply, environmental and health controls on the use of pesticides, fertilizers, food additives, and preservatives, price controls, and foreign sales policies, we discern different pieces of a grandly complex system. Working out this systems problem requires teamwork between government and private enterprise. Today the government has assumed a regulatory posture and no entity is providing leadership, so that the utilization of science is slowing down.

THE INSTITUTIONAL SECTORS OF TECHNOLOGY

Free enterprise acting alone can realize only a portion of the good that science and technology can bring us. The preceding examples suggest that government is a necessary participant in coping with the remainder. This is not a sanction for an enlargement of the governmental role in technology itself. There are some areas, such as defense and space exploration, where the government plays a leading role. But it should not seek to play a comparable role in the sponsorship of innovation where a private market exists. Capitalism may be unpopular, but even those who are vocal in criticizing it would probably distrust a massive entry by the federal government in resource allocation and industrial management.

Government and Industry

The federal government employs scientists and engineers in policy making and the management of technical efforts, including those of its own laboratories in specialized fields. But most of our nation's strength in science and technology is concentrated in private industry and the universities. If the government were to launch a major program of R & D with the avowed objective of raising the innovative technological level of the nation, it would have to influence, augment, and direct the efforts of thousands of trained individuals in the private sector and assume direct control of parts of the infrastructure serving them. Perhaps the worst outcome of such intervention would be to impair the market's contributions to the selection of product effort.

Technological advance is infinite in its variety and dimensions. New products that could be developed, the changes in methods and procedures that are technically feasible, and the possible alterations of virtually every system for producing, distributing, and using technical products and services are endless in number. The nation has only the resources, technically trained people, and funding to carry through a tiny percentage of all that can be conceived. The trick is to choose the right research and development, and to assign the limited resources, so as to meet the citizen's sense of values in the best way. Nothing about government suggests that it would be effective in sensing the market's real needs, that is, the true desires of consumers. Even those who are convinced that the free market is an inadequate means for matching resources to goals would probably concede that attempts by government to perform this balancing function would worsen the performance of the entire system. Only a small fraction of the electorate really believes that government could exercise control in this way.

In industrial technological innovation the government can help most, not by funding new efforts, but by improving tax policy and the general economy, so the private sector will have the financial means and incentives to perform more R & D of its own choice. Lowering capital-gains taxes and decreasing the double taxation on dividends would notably encourage investment in innovation. Germany and Japan enjoy higher investment ratios than ours and higher rates of productivity increase; their capital gains taxes are zero.

The Role of the University

The federal government must continue as the main sponsor of basic research in the nation's universities. In this role there are problems of performance. The funding level is low enough to leave good talent unused. The bureaucratic machinery allocates too large a part of the research dollar to the administrative expenses of universities. Also, the growing emphasis by government on proving the relevance of research underestimates the high payoff from probing unknown secrets of nature. We have gone beyond common sense in the process of control. It costs less to accept that some support will be useless than to try to guarantee that all research will pay off.

Universities also seem to be failing in their educational function if, as opinion polls indicate, many college graduates understand too little practical economics for responsible citizenship. It is too rarely understood, for example, that private industry must generate with its revenues the coverage of all costs, including that of capital. Too many college graduates are unaware that the United States ratio of investment to the total GNP is too low, especially by comparison with other industrial nations. Impressions of private industry profit margins are off by a factor of five, very commonly.

The creative and innovative process itself should be studied in the universities. In too few departments of institutions of higher learning do we find imaginative courses with the intention of stimulating and creating innovative talents among students. The relationship of certain social values, such as freedom, to technological advance needs to be better understood. If, as I am suggesting, organizational problems limit fullest use of science and technology on our behalf, then research should be directed toward national options to alter organizational structures.

Some Recommendations

Those interested in preserving the free enterprise system, such as leaders of businesses, should accept that ours is a hybrid society. It is not helpful and oftentimes harmful to press so hard as these supporters of the private sector often do, to have free enterprise labeled as the all-encompassing answer to all problems. Free enterprise, it should be understood, can provide only part of what is needed. A creative, cooperative partnership between the private sector and government should be sought, while adversary elements in their relationship should be played down.

Government leaders should recognize the involvement of government in science and technology as a permanent one and set a long-range policy to guide it. Such a policy should suggest answers to questions as to how the government and private sector should divide their support for areas of R & D. How can the government meet its regulatory responsibilities without damaging the national innovative effort? A healthy scientific effort should be counted by government as an indispensable national asset. Government should itself be innovative in fostering a climate for innovation.

Universities should mount increased efforts to discover and develop innovative talent. Their graduates should understand basic economic concepts so as to be able to function as citizen decision makers. Greater emphasis should be given to relationships among science, technology, and the social-economic-political structure of society. The universities should aspire to be sources of innovative ideas about innovation including alternative ways of structuring our institutions so as to enhance the possibilities of using science and technology to best advantage.

The public would do well to avoid extreme views. Free enterprise alone cannot do everything best. Nor is it so bad that government must do everything. If the public is interested in the flexibility, freedom of opportunity, motivation, and incentives of free enterprise, then it should recognize that industry must be allowed to generate the funds necessary for investment in R & D. And antitechnology sentiments should be set aside. Science and technology are not enemies without but the public's own tools. Citizens should strive toward their use in attaining goals they have chosen.

8
Public Reactions to Science and Technology: "The Wizard" Faces Social Judgment
Jean-Jacques Salomon

> Lorsqu'il s'agit d'une tentative nouvelle, devant quoi reculcerait un physicien? l'existence d'autrui? la sienne? Ah! quel savant, digne de ce titre, pourrait, ne fût-ce qu'une seconde, songer, sans remords et même sans déshonneur, à des préoccupations de cet ordre lorsqu'il s'agit d'une découverte? Edison, à coup sûr moins que tout autre, Dieu merci!
>
> Villiers de L'Isle-Adam in *L'Eve Future*[1]

The myth that grew up around Edison in his lifetime is of considerable interest today. It illustrates with great clarity the ways in which public reactions to science and technology have changed since Edison's time and above all in the last quarter of a century. Beyond the legend of the genius, the inventor who was forever able to "make things," the myth of Edison celebrated an alliance of science and industry which seemed destined to assure the triumph of the ideal of the Enlightenment, technical progress conceived as the instrument and guarantee of social and also of moral progress. Yet it is this very union of science and industry—a union which was consummated ultimately through the intermediation of the state—which has made suspect the benefits of scientific and technical research.

It is a characteristic of myths that they endure in spite of the counterevidence provided by reality and by historical research. The myth of Edison is far from being an exception to this rule. It indeed has an almost exemplary value; it expresses the need of every society to find itself in a self-image which justifies its own most profound attributes. In the second half of the nineteenth century, an epoch that consecrated the power of industrial capitalism, Edison was more than simply a magnificent example of an inventor and engineer who succeeded. He was the model of an entrepreneur who vaunted materialist values sanctioned by commercial success, as against the theoretical scientists, and the intellectuals more generally, whom he accused of being unconcerned to find uses for their researches.

In his saga of industrial power and the American dream John dos Passos relates the legend of the "Electrical Wizard" with a conciseness—in three and a half pages—in which all is said: the self-taught man, the lone-wolf inventor, the tinkerer who made a business of invention itself. In one phrase, Dos Passos expresses the reasons that Edison's genius evoked a myth which fascinated a public concerned more with the prestige of material success than with intellectual values. "He never worried about mathematics or the social system or generalized philosophical concepts."[2]

As to the first point, Edison's apparent disdain for mathematics, his biographers as well as historians of technology have demonstrated capably that the public's fascination reflected a misunderstanding of the inventor's methods. Yet it is still worth coming back to this popular misconception, for it is closely related to the two other points, disdain for the social system and for generalized philosophical concepts. There is a link between the former and the latter which has to do with the spirit of the period, with a conception of the social role of the scientist and with an attitude of "technological laissez-faire" which in turn sustained the myth of the uncultivated but genius inventor, capable of achieving anything and achieving everything of which he was capable. It is precisely this attitude, this conception, this spirit, which the public today tends to doubt, even to repudiate.

EDISON NEVER WORRIED ABOUT MATHEMATICS

A self-educated man, a former trainboy, who took out so many patents and produced so many inventions deserves to become a legend. Edison was indeed not so much exceptional as unique. His legend reflects the traditional idea of invention as a divine gift, the pure expression of a genius inspired only by

Providence, in the sense in which Alexander Pope wrote of Newton, "God said, let Newton be, and all was light!" But Edison was not simply an inventor, in which case he would belong in the Musée Lépine rather than in museums of science and technology. What he represents is not so much the end of the heroic age of invention as the beginning of the scientific age of technology.

Edison was a self-taught man, but so were such predecessors as Davy and Faraday. All his life he read a phenomenal quantity of scientific books and journals, which he used to further researches of his own. At the age of fifteen he was repulsed by a cursory reading of Newton's *Principia*, which gave him, in his words, "a distaste for mathematics from which I have never recovered". Yet the decisive event in his career as a researcher came when he read the two volumes of Faraday's *Experimental Researches in Electricity*. Aged twenty-one, he found his bible of experimentation and a revelation as to the ways in which laboratory work, without benefit of complex theory, can illuminate an understanding of natural laws.

It is true that Edison lacked the power of abstraction which provides access to mathematics. But one cannot conclude that his disdain for mathematics led him to make his inventions without recourse to the disciplines of science (chemistry in particular). If Menlo Park was the first modern industrial laboratory, it was not simply because it was subsidized by the Morgan-Vanderbilt syndicate in order to complete a particular research program, but above all because it drew on a staff of well-trained scientists and on the most advanced scientific equipment. Without Francis R. Upton, a specialist in mathematical physics trained at Princeton and in Germany in von Helmholtz's laboratory, the technical problems of voltage and resistance posed by the development of the incandescent light would never have been resolved. The famous formula, "I can hire mathematicians but they can't hire me," should not hide the fact that Upton's equations played a decisive role in many of Edison's inventions, as did the machines and scientific instruments Edison had brought to Menlo Park—galvanometers, static generators, Leyden jars, induction coils, batteries, condensers—without which the realization of his ideas would have been impossible. The best example is that of the vacuum, indispensable for the perfecting of the light bulb. It was from his reading of scientific journals and proceedings that Edison became aware of Sir William Crooke's achievements with high vacua, thanks to the Sprengel pump. The moment he learned that one of these Sprengel pumps was to be found in Princeton he borrowed it and from then on Menlo Park had available the latest vacuum pumps. It should be added that the library at Menlo Park was so rich in scientific and technical books and periodicals, domestic and foreign, as to rival the bookshelves of the best departments or university institutes of the period.

Is it necessary to recall how little was known at the time about electricity, and in particular about the transmission of current? Progress could only come about from a two-way relationship—a "cross-fertilization" as we would say today—between theoretical principles and practical applications. In seeking to "subdivide" a current so that each lamp could be used independently, Edison attacked a problem which theoretical scientists believed impossible to resolve. According to such authorities as Lord Kelvin and John Tyndall the laws of electrical resistance denied that a light with a very high resistance could use little current. "Knowing something of the practical problem," Tyndall said, "I should certainly prefer seeing it in Mr. Edison's hands to having it in mine."[3] The solution, which Edison glimpsed very soon, was as simple as it was brilliant—the use of parallel circuits at high voltage so that an incandescent burner with a small cross section could be used.

But if it was Edison rather than the theorists who had the idea of this solution, its application and perfection depended on the cooperation of the entire team at Menlo Park, including mathematically trained men such as Upton, Andrews, Clarke, Acheson, Sprague, all university-trained scientists who surpassed Edison in theoretical knowledge; technical men such as Boehm, an artistic glass blower who had been trained in the famous Geissler works in Germany; collaborators from his first telegraph and stock ticker machine manufacturing shop, Batchelor, the mechanical draftsman, and Kruesi, the former Swiss clockmaker; and Ott, Bergman, and Schuckert, who could construct almost any form of instrument or machine. "The Wizard" was indeed the conductor of the first laboratory organized for industrial research, founded on theory and at the same time on practice, "a charismatic leader of team research . . . who was more than an inventor, researcher and engineer—he organized and presided over scientific and technological enterprise."[4] Far from being a simple inventor, or from depending on Providence or luck, he paved the way to the systematic research of the technological age, with a whole team of researchers comprising machinists, technical men, scientists, and other workers. Menlo Park, his "invention factory," may have been, as well, his greatest invention.

The hold which Edison exercised on the fascination of the general public was equalled only by the reluctance of established scientists, trained in universities, to look beyond the "tinkerer" capable of resolving technical problems so as to acknowledge his strictly scientific findings and intuitions. When he worked in 1875 on the problems of "acoustical telegraphy," Edison discovered that scintillating sparks issuing from the core of a magnet might be caused by "something more than induction"—an "unknown non-electric force," to which he gave the name of "etheric current" Edison's observations were challenged and even ridiculed by learned representatives of academic physical science, Professors Elihu Thomson and E.J. Houston in Philadelphia, and Silvanus Thomson in London, in such a manner that his prejudice against theoretical scientists—as distinct from theoretical science per se—never ceased. These etheric sparks were considered "spurious." Although Edison was wrong in concluding that the sparks were "non-electric," by producing high-frequency electric waves he had been on the threshold of electronic science.

But this contretemps before a well-established audience did not stop after his death. Hughes has shown beautifully how the tributes paid in *Science* in January 1932 by representatives of the new generation of scientists—Frank B. Jewett,

Robert A. Millikan, Karl T. Compton—reinforced the image of Edison as a folk hero rather than acknowledging his pioneering role as a founder of industrial science. Whatever his heroic dimensions as an inventor and an engineer, he could not measure up against the new standards of industrial science nor even against the definition of the new technologists produced, such as Millikan at California Institute of Technology or Compton at Massachusetts Institute of Technology, under the authoritative guidance of academic science. Hughes explains the bias of these critics by suggesting that they "realized that until Edison and his method were believed antiquated, the possibility remained that pragmatic corporation executives might ask why young Ph.D.'s from the universities were being hired at substantial salaries to do what Edison and his type were known to have done so well—introduce fantastically remunerative inventions."[5]

This clash of styles—formal education and modes of publication, but also language and even dress—between Edison and the new generation may have to do with differences in scale, specialization, complexity and wealth of industrial science in the twentieth century as contrasted with that of the nineteenth. Still, as Hughes writes, the representatives of the new generation "should look to the Edisonian method for a better understanding of themselves." Through his notebooks and the way in which he solved technical problems by using the laws of Joule, Ohm, and others, the evidence is abundant that Edison's work was indeed "oriented research" that used science. But the stereotype image of the tinkerer who "never worried about mathematics" reinforced by the new generation of scientists suited too well the expectation of the American public who needed to believe in the "technological hero."

It is fair to add that Edison himself enjoyed being the embodiment of this popular myth. Quite aware of the classic American values which he epitomized—individualism, materialism, do-it-yourself solutions and basically anti-intellectualism—he was keen to cultivate a popular image that matched these values. Here comes the empirical crusader of invention who at the same time as he improvises any technical solution makes a fortune from it! Undoubtedly, the legend of "the Wizard" was also a product of Edison's gifts as an actor, a showman, an advertiser—a mixture of Dr. Faustus, Barnum, and Henry Ford. But the expectations, hopes, dreams and myths raised by science and technology in the nineteenth century were such that Edison could appear as its very symbol, that of an endless technical progress to which everything is possible and everything possible has to be done *without* mathematics or theoretical science. This coincidence between invention and simplicity, fooling around with matter as if matter can never resist the attack of craftsmanship, should necessarily result both in the fortune of the individual innovator and the prosperity of mankind.

EDISON NEVER WORRIED ABOUT THE SOCIAL SYSTEM

Fascination, adulation and even canonization,[6] the mythic ideal of the inventor diffused by Edison "as part Tom Swift, part hermit,"[7] flowed as direct consequences from two beliefs. First, since he was simply a craftsman and not a man of science any average man (especially American) could dream of emulating him. Second, his inventions epitomized the inescapable usefulness and beneficence of the industrial arts in the new technological era. As to the former point, the popular mind was basically mystified. As to the latter, what happened in the twentieth century as a result of technological proliferation has challenged the optimism associated with the "miracle-making process" of technical progress. It seems to me that, if the panegyric of science and technology in our days is not as unconditional as it was in Edison's more positive time, this change has *also* something to do with what was his true method of invention.

"The End Justifies the Means"

Edison's better biographers emphasize that his inventions, far from being mere devices, were components in whole technical and economic systems.[8] Most of his inventions illustrate this method, which was not all that of an unlettered tinkerer searching and trying by inspiration and persistence, but that of a professional inventor and entrepreneur conceiving research as a way of synthesizing the technical, scientific, and economic aspects involved in the process which leads from invention to innovation. The quadruplex telegraph, the telephone, the phonograph, magnetic ore separation, the storage battery—all these examples show in different ways how Edison presided over the entire process. But the story of the incandescent lamp invention is certainly the clearest and most convincing example.

In a sense, the lamp was his simplest achievement, a small component in the integrated system of lighting which implied not only many other components— "parallel" distribution, jumbo generators, mains, underground conductors, junction boxes, fuses, all of which (except the boilers and steam engines) had to be conceived and developed by Edison and his team—but also the need to construct all parts with reference to all others, the interactions within the system leading in turn to new inventions and patents. If Edison defined invention by his famous formula, "99 percent perspiration and one percent inspiration," it was also because he could not see the process of finding and inventing culminating in its technical phases; these were never distinct from their economic implications. For instance, as Hughes has underlined, the decision to seek a high-resistance lamp filament in contrast to the low-resistance one generally tried (and which was considered by learned professors of physics to be the only solution possible) "was a logical deduction of cost analysis" rather than the result of ingenious experimentation and sustained reasoning.[9]

If the incandescent light was to be economically competitive with gas, the cross section of the copper conductors had to be increased to reduce loss in distribution, but this would enormously increase copper costs; therefore it was necessary to reduce current and to still have enough of it for lighting the incandescents. The application of Joule's and Ohm's laws (raising the voltage in relation to the current) was the technical and indeed scientific breakthrough required to solve the economic problem. From then on started the problems of developing and producing the system on a real scale, from the lighting of Menlo Park to the demonstration of Pearl Street, and this again was a blend of careful technical solutions and extremely close economic calculations. Edison was not tinkering with gadgets, he was facing a technical-economic challenge the answer to which was not merely the electric light, but the electric home, shop, and factory.

His flashes of intuition, insight, and experimental gifts cannot be understood if they are not placed within the context of both a certain age of science and of capitalism. The legends of "the Wizard" appealed all the more to the popular mind because, in Josephson's phrase, "Edison had a *Homo oeconomicus* within him."[10] In the triumphant Industrial Revolution, the inventor as folk hero was also a hero of the capitalist wonderland in which the marriage of technical skills and dollars was to make progress the "endless frontier." The child born from this marriage has no other name than modern technology.

This view of Edison as a *Homo oeconomicus* implied in nineteenth-century capitalist doctrines that his commercial faculties served a well-developed social sense. But what kind of social sense? Successful inventions meant new jobs and greater welfare so that their usefulness contributed to making economic growth equivalent to social progress. In the process of developing and producing these inventions on a large scale, there were, of course, some side effects, but such was the technological spread, like Bergson's *élan vital*, that it could only proceed for the betterment of mankind. Indeed, Edison's inventions had already led to such side effects and accidents due to the crude wiring and insulation of the light system. The Pearl Street station caused leakages and shocks to men and horses on the pavement above the conductors. Mrs. Vanderbilt's house almost burned from the crossing of two wires. Even worse, a customer was killed by a destructive fire in a house wired by Edison's mechanics. But this was a small price to be paid in tribute to progress.

More revealing of the forthcoming wilderness of an uncontrolled rise of technology was the dispute over the advantages of the alternating current introduced by the German engineers, adopted by his rivals Westinghouse and Elihu Thomson, and opposed (wrongly) by Edison on the ground that it would be too dangerous to work and live with. In order to make his case, Edison organized public sessions of electrocutions of stray cats and dogs by means of high-tension currents. "The feline and canine pets of the West Orange neighborhood were purchased from eager schoolboys at twenty-five cents each and executed in such numbers that the local animal population stood in danger of being decimated."[11] These lugubrious if not sadistic experiments did not help Edison to demonstrate that his DC system was safer than the AC system; at least they led one of his former assistants, H.P. Brown, to propose the electrocution of criminals sentenced to death by charges of alternating current—as he said, "an instantaneous, painless and humane death." The substitution of the "electric chair" for hanging, adopted by the New York State Legislature in 1888, was also a proof of progress. The French Revolution had led to the mechanical guillotine; the Industrial Revolution could rely upon "fairy electricity." Still, in this "war of currents," Edison, trying to persuade the public that the AC system was fit only for mad dogs or criminals, went so far as to declare, "I have not failed to seek practical demonstration. . . . I have taken life—not human life—*in the belief that the end justified the means*".[12]

The End of Innocence

It is at this point that one can see to what extent public reactions to science and technology have changed in one century or less. The crux of this controversy over the two currents was technical and certainly economic; but neither the adepts of the DC Edison system nor those of the AC Westinghouse system "worried about the social"—human and moral—implications at stake. The value of a technical system is not that it satisfies human needs; it is that it works more efficiently and economically than another system. It is from this standpoint that "the ends justify the means," whatever may be the social costs. This is what I have called the "complex of technology for its own sake," or in J.R. Oppenheimer's words, the "technically sweet" solution which makes it imperative to realize a project once it has been deemed feasible, no matter what the consequences.[13] This "complex" leads to the suppression of everything which links the problem raised and its solution to the human and social environment and gives it significance in relation to that environment. The assertion of the "neutrality" of technology (and science) in this context assumes its most aggressively deceptive form, as though the esthetic or economic value of the technical solutions had nothing to do with the use of the tools which are forged with their aid—in brief, as though there were no relation at all between science, technology, and values.[14]

Indeed, as a result of the union of science, technology and industrial capital, something happened in the twentieth century which now restrains science and technology from asserting the quality they proclaimed in the nineteenth century—in a word, their *innocence*. It is not only that the age of invention has become incorporated and capital intensive, leading to innovations so thick and fast that they directly affect social change and individuals' lives on a scale and with a complexity which seem to challenge any rational control. It is also that science itself can no more be isolated from technology. Just as the frontier between pure science and applied research has become blurred, the frontier between knowledge and its application no longer really exists. And last but not least, it is that science and technology can no more be dissociated from political power. Just as the increase and the influence of research activities meet the needs of the political system, state priorities meet the needs of science and technology.

From this change, one cannot conclude that the public is completely disillusioned with science and technology. The evidence from public opinion polls and attitude surveys indicates that scientists and engineers are held in a higher regard than most other professions and occupations and that the man in the street still expects many benefits from scientific and technological research. The problem is not that the faith in knowledge tends to be replaced by a bet on ignorance and obscurantism; it is that there is a rising level of mistrust as to the ways in which science and technology influence social change, and more and more evidence—in the media as well as in the political arena—of public concern about certain of the consequences of technological development.

This change may be said to date back to World War II with its symbolically important events such as the Manhattan District Project, the first nuclear explosion at Alamogordo, or those at Hiroshima and Nagasaki. But it could just as easily be traced back to World War I which, as André Malraux pointed out, showed for the first time, through the use of poison gas, "the adverse side of the science balance sheet". Since World War II, and now in the civilian as well as the military sphere, there has been a rapid succession of such symbolically critical events—Minamata, Thalidomide, Seveso, Torrey Canyon and Amoco-Cadiz—which give the critics and protesters all the more justification for denouncing this change of course as a betrayal. Associated with the threat of the atom, the growing sophistication of so-called conventional wars, the deterioration of natural and social environment, the quantitative demands of economic growth and the consumer society's boundless aspirations, the scientific enterprise finds itself attacked not only from *outside*, but also from *inside* the scientific community. The image of the researchers—scientists as well as technologists—as wizards making magic that leads necessarily to more bountiful life, has turned into a new priesthood allied with industrial, political and military power and making magic which appears less and less white.[15]

Technical Progress on Trial

One of the causes of malaise in our societies since the 1960s is certainly linked to the awareness of the damage done by the growth process as conceived up to now. It is far from being the only cause, but it is a particularly significant example, for two reasons. First, it underlines that growth is not without limits and second, that the industrialized societies, in estimating gain and loss, have to take both into account. In this process of "creative destruction," as Joseph Schumpeter called it, it is no longer possible to gloss over what is destroyed. In addition, this example reveals a totally new attitude of society toward science. Since science cannot disclaim responsibility for the damage done, it is no longer seen as synonymous with progress. Until very recently, scientific activity could blossom without any danger of being shown up as the culprit or accomplice responsible for disastrous consequences; today, it is almost obliged to prove its innocence in advance. Such is the underlying meaning of "technology assessment." Certainly the term implies evaluation and control of technology, but etymologically the formula also refers, both in French and in English, to the idea of a court of justice. Now not only technology but also science must appear before this court. This is obvious in looking at the problems raised by recombinant DNA and the threats inherent in the possible manipulation of genetic capital. From both inside and outside the scientific community, research is suspect as if, for the first time, reason no longer trusted reason, and proposed that the fantastic powers of the wizards should be curbed.

In speaking of recombinant DNA, Professor John Kendrew recalled the two parables of the "Genie and the Bottle," one in the *Arabian Nights*, where the Genie reveals himself as good or bad depending on whose hands release him; and the other about the old fisherman who, when the Genie threatens to kill him, shuts the Genie up in his bottle and, before setting him free again, makes him promise to let him live. The first, the parable of Aladdin, refers to the classic theme of good and evil use of science; the second raises the more topical problem of control over scientific power. After going over the potential dangers of recombinant DNA, however, John Kendrew concluded that there is no reason to suppose that scientists are "stupid or irresponsible enough" to follow the worse direction.[16]

Such an expression of confidence seems to me merely an article of faith, for, it is no longer enough to trust in reason, because this century has taught us—and at what price!—that reason does not give birth to monsters exclusively in its sleep, as Goya suggested. Conscious reason can also become a nightmare. For all large-scale innovations, the question still remains beyond what level does progress become a disadvantage or a tragedy; and since the Genie is already out of the bottle and there is no chance of shutting him up again, is it still possible to keep him on a leash?

True, it is hard to see how any moratorium or other limits could be imposed on scientific investigation.[17] But the price to be paid if there is no control is obvious. The lesson of armaments and nuclear weapons makes it impossible to be content with the gospel ingenuously proposed by scientific optimism. We know that escalation is endless, that any breakthrough here leads to another, and that there is not, that there never will be, any adequate *technical* solution to stop it. On the contrary, technical breakthroughs continually endanger the very security they are supposed, traditionally, to establish, so that today the preparation for war is synonymous with the impossibility of peace.

Scientism in the nineteenth century considered scientific and technical activity as the most decisive factor in human progress. The combination of social aspirations with the development of logic peculiar to science and technology was supposed to make possible, through one change after another, the reign of a better social order. Opening up by the same token the mastery of social relationships and those between society and nature, more and more efficient technology would ensure that progress would continue. But today this optimism is called in question; the changes due to scientific and technological development are not necessarily factors of progress. Science and technology, objects of both hope and fear, appear more and more visibly as the reflection and the catalyst which reveal society's major problems; they have laid themselves open to social conflicts by becoming one of their stakes. Thus it is insufficient that the scientific and technological entrepreneur has a *Homo oeconomicus* within him; the fears and threats associated with technological development—as they are more and more reflected in public opinion and legislation—compel him to have as well a *Homo sociologicus* and even *ethicus* within him.

EDISON NEVER WORRIED ABOUT GENERALIZED PHILOSOPHICAL CONCEPTS

The current unease in regard to science and technology is obviously to be related to the speed, scale, and complexity of the changes determined in our days by modern technology, as well as to the cumulative, universal, and more often than not irreversible character of its effects on the human, social, and natural environment. The question posed in the *EPRI Journal* for October of 1978 in the announcement of the Edison Centennial Symposium, "Science and Technology—Gone Far Enough?" could easily have been, "Science and Technology—Gone Too Far?" Yet, this malaise is not a new thing, although it started becoming more acute after World War II and even more in the turbulence of the 1960s.

Science and technology—gone too far or gone with a wrong conception? Our societies are spellbound by the weight and the prestige of science and technology to the point where it is no longer possible to find any place in the world unaffected by this spell. The triumph—or trickery—of reason ends up giving the irrational the guise of rationality itself. This century is also the age of the absolute threat of the atom bomb; of the nonsense of economic growth conceived as an end in itself; of the mockery of the consumption of the advanced societies as compared with that of the two-thirds of the world societies, subject to poverty, malnutrition, and famine; and now, within the most industrialized countries, a yearly average of 15 million unemployed people. "One cannot stop progress," but progress to what end? Plato's lesson appears to have been forgotten. What is the use of the science of navigation if one does not know where to go?

The Failure of Humanism

"The exclusiveness with which the total world-view of modern man, in the second half of the nineteenth century allowed itself to be determined by the positive sciences," charged the German philosopher Edmund Husserl, "and to be blinded by the prosperity they produced, meant an indifferent turning away from the questions which are decisive for a genuine humanity." He employed the word "prosperity" in English in his original text. Husserl denounced this "failure of humanism" as early as 1935, when barbarism was spreading inexplicably across Europe. "We know that it has gradually become a feeling of hostility among the younger generation. In our vital need—so we're told—this science has nothing to say to us. It excludes in principle precisely the question which man, given over in our unhappy times to the most portentous upheavals, finds the most burning."[18]

If science is rebelled against, even by those in its service, it is not because "it has not gone far enough" nor even because "it has gone too far"; it is because it appears unconcerned with value problems. Still, beyond the difficulties to which it has given rise or which it has failed to solve, one has the feeling that the present revolt against it is directed not merely against the ends it has served, but also against the principles and internal procedures underlining it. This is to confuse the misdeeds of a social practice of science subject to the overriding pressures of the industrial, political and military system, with some evil inherent in the scientific method as such. But the failure of a *certain form* of rationalism is not the failure of *all* rationalism. As Husserl said, "the reason for the failure of a rational culture lies not in the essence of rationalism, but in its entanglement in objectivism"—in other words, in the exclusive domination of the natural sciences over our culture. In this sense, the current crisis is primarily the crisis of the "two cultures" which we have been unable to persuade to live together in harmony.[19]

The distinction between the sciences and the arts in education is something that arose roughly with the Industrial Revolution, and so may be a product of a particular phase in our societies rather than some God-given necessity. Certainly, this divorce was indispensable in the eighteenth century in order to sever the bonds between theology and the natural sciences. In the nineteenth century, specialization and professionalization among researchers were soon to consolidate a new intellectual community whose work, norms, language, and channels of communication had to be, in order to make progress, under its own control. But the exercise of such a control over the scientific method did not imply that the scientists, interested in a smaller and smaller sector of reality and an even smaller and smaller portion of science, had to disengage themselves from the rest. Even Auguste Comte, the founding father of positivism, could never say enough against his scientific colleagues who separated science from philosophy. When, as he wrote, "they give preponderance to the spirit of the detail over the spirit of the whole" that just proves the "philosophical impotence now proper to our learned societies . . . [which results in] the political impotence of the scientific class and even its moral degeneration."[20]

I have tried to show how the substitution of the "scientist"—a word coined by Whewell only as of 1840—for the *savant*, the "learned man" or the "natural philosopher"—resulted from a cultural remodeling: the researcher, distinguishing himself from the man of learning or the teacher, allied himself with the producer; the death of the *savant* led to the apotheosis of the scientist. The former tended to transcend technical knowledge in a moral commitment aiming at something more than the mastery of a speciality; the latter claims that he is only performing value-free acts and that he is unskilled in whatever intellectual problem does not hold to his technical knowledge. And so the scientist became a technician among other production agents as contrasted to the *savant* who functioned in the ideology of our societies as a *cultural figure*.[21] This predicament of technical value-free specialization, concerned only with efficiency and useful things, is directly related to the disdain for the liberal arts and the humanities. Nobody expressed it better than Edison when he was interviewed about formal education. "What we need are men capable of doing work. I wouldn't give a penny for the ordinary college graduate, except to those from institutes of technology. . . . They aren't filled up with Latin, philosophy, and all that ninny stuff. America needs practical, skilled engineers, business managers, and industrial men. In three or four centuries, when the country is settled and commercialism is diminished, there will be time for literary men."[22]

The Challenge of Education and Participation

There are at least two levels on which action is needed to restore communication and better understanding between the scientific and technical community, many members of which are as disillusioned as the general public, and the man in the street as well as society as a whole. The first embraces precisely the very conception of education and training of the technologists—scientists and engineers. Our modern education is split into two heterogeneous worlds, an "active" continent, embodying the spirit and method of science, and a "humanist" island, which is not acknowledged to have any practical value. This "humanist" culture is negatively defined since it is not supposed to act on things and is at the same time depicted as "backward looking," rejecting the present and the future.

Even the expressions "hard sciences" and "soft sciences," which express in terms of weight the intellectual imbalance between the various scientific disciplines, speak volumes about our prejudices. But still more revealing is the lack of reference to any other branches of knowledge than those enshrined in the university curriculum—as if our modern thinking stopped short at this duel between the giants and the dwarfs, the natural sciences on the one hand, and the social sciences and humanities, on the other. If the confrontation of the two cultures is a drama, it is not, as C.P. Snow believed, because they reflect two rival sets of tastes, aptitudes, or skills, but because they represent *opposite functions*, which the industrial system chooses to regard as irreconcilable.[23] A good example of this is given by Noble when he describes how engineers are educated and molded so that they strictly serve corporate ends.[24]

The scale of values of the industrial society gives precedence to those aptitudes and activities which most effectively satisfy its needs, but the technician, with an almost exclusively scientific training is, more often than not, as handicapped in the world of life and values as is the person with no scientific background in the world of technology. As long as the educational system cannot find more room for teaching the future scientists and engineers in subjects outside the sphere of the "hard sciences," the schism between the cultures is bound to deepen, setting technicians unable to understand value problems in opposition to others baffled by the esoteric technicalities of science.

Then there is another technical level of action which depends both on the community and on public authorities. Scientists and engineers are the first to draw society's attention to the implications of their work. To be sure, the notion of the scientist's social responsibility contrasts in a large measure with the traditional concept of scientific responsibility. To give a warning to society in the early stages of research is, in fact, to go beyond "neutrality" and "objectivity"; it implies interpreting preliminary results in the light of theories which will perhaps not stand up. If these theories turn out to be mistaken, is there not a danger of unnecessary alarmism? But to remain silent would be to commit an even graver offense; scientists and engineers do no more than call society's attention to a possibility; it is not up to them to take a decision that remains fundamentally political. Already, various scientific organizations have been concerned with these problems, or have been set up with this aim—for example, the Council for Science and Society in England or the "Mouvement Universel pour la Responsabilité Scientifique" in France. The new scientific ethic which is at present being worked out will make it easier to develop the public debate which is essential if society is to be ready to solve the problems raised by science and technology.

This suggests the need for a new kind of dialogue between society and researchers. This dialogue would take into account the aspirations of all representatives of a now highly complex social system, rather than being limited to science and technology's own trends and pressures. We need to ask what science and technology are aiming at, what kind of society. Have we not lost sight of the aim because scientists, engineers, decision makers, and the general public all firmly believed that the technology which had come out of almost autonomous scientific and industrial developments was a good thing in itself?

To speak of a social assessment of technology does not mean to expect that technical expertise will provide the apparently objective advice which would avoid the disadvantages of progress. It means, above all, consulting all the elements of society especially liable to suffer from these disadvantages. In this sense, research activities are not and can no longer be the exclusive business of researchers or the institutions which subsidize them. Since consumers, trade unions, and local communities are all concerned by the positive or negative effects of technical change, it is imperative that they exercise the right to look at the way in which research activities conduct this change. If science policies are to be effective, the scientific and technological enterprise must both come out into the open and be better integrated in society. This is obvious from the fears caused by the implantation of nuclear power stations; as long as there is not closer public participation in such decisions, technological development will appear as a threat rather than a solution. But nuclear power stations are only one example among many where the large-scale diffusion of technologies, with their effects on both natural and human environment, causes public opinion to wonder about the "negative" consequences of technical change and leads to the establishment of more and more groups to combat these consequences.[25] Certainly, nuclear energy adds one more dimension to the supposed or real disadvantages of technical progress, fear of the atom. Probably nothing will conquer this fear, although the dangers of nuclear power stations are as nothing compared to those of nuclear weapons, of which public opinion is less aware. The "original sin" which, to use Robert Oppenheimer's expression, stamped the mastery of fission, whose first use was destruction, can never be kept out of the debate.

Whatever the technologies involved, we have to wonder whether there are ever any real "experts"; or, more precisely, whether any incontrovertibly demonstrated expertise can exist. The scientist's authority is based on the idea that his interpretations, estimations, and forecasts are above the conflicts about values and interests which are involved in any political discussion, because they are based on objective data, methodically collected and analyzed. This supposed rationality and objectivity gives the scientist's opinion the power to depoliticize a controversy. And it is no accident that the growing recourse to expertise in the 1960s, precisely because of the multiplicity and complexity of the technical questions governments had to settle, led to the belief that here was an "end to ideologies" and that the days of "irresponsible" crusades by incompetent militants and philosophers were numbered, since these questions were to be debated on the uncontested ground of scientific discourse. In fact, exactly the opposite happened. Far from confirming that decisions were rational, expertise ended up by looking like an operation of wizards where "those who know" had to be believed, simply because they were the only ones who possessed technical knowledge on a specialized subject. Far from doing away with ideological battles, recourse to experts made them rebound, not only because the experts' objectivity is all the less to be taken for granted since they may be divided among themselves and question each other's judgment, but also because the issues at stake cannot be reduced merely to the technical data which make up the debate.

It can at least be seen in this way how the rhythm, nature, and scale of technological change saddle the industrial societies with a double duty, to be more careful how they use new techniques, if they do not want them to end in catastrophe, and to take more account of public opinion, if they do not want to become paralyzed. Since decisions have to be taken in any case, the only way to implement them is to get them accepted—if not shared—through full information and, wherever possible, participation of the public in the decision, by giving access to technical data and, above all, by presenting these data as what they are: elements of a case which will be legitimate not because of its technical characteristics but only through a political choice.

Criticism of the scientific and technological enterprise is not just one of the ups and downs of social criticism (or self-criticism) which rationalism has always undergone. Until the nineteenth century, when the natural sciences were set up into separate branches of both theology and philosophy, social criticism in fact based itself on the intellectual and practical achievements of science in order to denounce the injustices and mistakes of the existing order. Today, these very achievements have led to a contesting of science itself, and this dispute is at the same time the most revealing and the most contradictory element in our present crisis; the most revealing, because the institution which embodies the success of rationality in all its glory is challenging its own tenets, the most contradictory, because this crisis cannot be overcome if it means a farewell to the opportunities still provided by science and technology.

Between innocence lost and an uncertain future, between utopia and the constraints of reality, it is clear that what is really at stake is the discovery of one or more new paradigms not only for science and technology, but also for society as a whole. Yet turning our back on scientific processes or refusing technological resources will not give us the least chance to make good the damage caused by progress or to overcome future dangers. If science and technology are our destiny we can, we must, learn to make better use of them. This implies that scientists and engineers must face the value problems involved in what they do and that the wizards of our technological era must worry even about "generalized philosophical problems."

NOTES

[1] Villiers de L'Isle-Adam, *L'Eve Future* (Paris: Jean-Jacques Pauvert, 1960), p. 31 (a science-fiction novel inspired by Edison).
[2] John Dos Passos, *The 42nd Parallel* (New York: New American Library, 1958), pp. 308-311.
[3] Matthew Josephson, *Edison—A Biography* (New York: McGraw-Hill, 1959), pp. 197, 479, 181, 347, 349, 440; and F.L. Dyer and T.C. Martin, *Edison: His Life and Inventions*, 2 vols. (New York: Harper and Brothers, 1930).
[4] Thomas P. Hughes, "Edison's Method," in *Technology at the Turning Point*, William B. Pickett, ed. (San Francisco Press, 1977), p. 5
[5] Ibid.
[6] One has only to think of the Roman triumph *à l'américaine* which was organized for the fiftieth anniversary of the invention of the incandescent lamp. A fantastic delegation headed by President and Mrs. Hoover, including the nation's most eminent political, industrial, and financial personages (and some scientists such as Marie Curie) "travelled in a little train of the Abraham Lincoln era to the restored 'Smith's Creek Junction' where, seventy years before, Edison was supposed to have been given the bounce out of his baggage-car laboratory." Like the young trainboy who used to sell newspapers, the old man walked about for a few moments with a basket of merchandise, crying "Candy, apples, sandwiches, newspapers" and offering them to the president. Josephson adds: "It was absurd, and yet also sheer symbolic drama; the American Dream reenacted before the world's newspapers and movie cameras" (Josephson, *Edison—A Biography*).

⁷Gene I. Rochlin, ed., *Science, Technology and Social Change: Readings from Scientific American* (San Francisco: W.H. Freeman, 1974), p. 104.

⁸Josephson, *Edison—A Bibilography*.

⁹Thomas P. Hughes, "The Electrification of America: The System Builders," mimeographed by the author, 1978.

¹⁰Josephson, *Edison—A Biography*.

¹¹Ibid.

¹²Ibid.

¹³In the controversy between Oppenheimer and Teller over the possibility of building the H-bomb, Oppenheimer admitted that his conversion to the project was solely due to the changes in the technical conditions for its construction: "The program in 1951 was technically so sweet that you could not argue about it. It was purely the military, the political and the human problems of what you were going to do about it once you had it" [*In the Matter of J.R. Oppenheimer: A Treatment of Hearings before the Personnel Security Board* (Washington, D.C.: U.S. Government Printing Office, 1954), vol. 1, p. 251.].

¹⁴Jean-Jacques Salomon, *Science and Politics* (Cambridge, Mass.: Massachusetts Institute of Technology Press, 1973) (first published in Paris: Seuil, 1970).

¹⁵See the recent comment by Harvey Brooks: "The scientists of my generation look upon the 30 years since the war, but especially the decade of the 1960s, as a golden age for science and for the flowing of scientific technology. Yet, for all of us, it also stands in the shadow of the most intense application of intellectual talent and theoretical knowledge to the arts of war in the history of mankind. . . . It was true that they [the scientists] were acting out of what they believed at the time to be the most worthy and constructive of motives, but like others they were prisoners of the political climate of their times, and their views, as in other cases, may have been colored by an element of rationalization of personal ambition. It is only too easy to apply the wisdom of 20/20 hindsight to their activities" [Harvey Brooks, "Technology: Hope or Catastrophe?" *Technology in Society* 1 (Spring, 1979), p. 7].

¹⁶John Kendrew, *Recombinant DNA: The Genie and the Bottle?* Lecture to the General Assembly of the European Science Foundation, Strasbourg, 1977.

¹⁷Robert S. Morison, ed., *Limits of Scientific Inquiry, Daedalus* (Spring, 1978).

¹⁸Edmund Husserl, *The Crisis of European Science and Transcendental Phenomenology* (Evanston, Ill.: Northwestern University Press, 1970), pp. 5-6.

¹⁹Ibid.

²⁰Auguste Comte, *Cours de Philosophie positive* (Paris: Schleicher Frères, 1908), p. 273.

²¹Salomon, *Science and Politics*.

²²Josephson, *Edison—A Biography*.

²³C.P. Snow, *The Two Cultures* (Cambridge: Cambridge University Press, 1959).

²⁴David F. Noble, *America by Design—Science, Technology, and the Rise of Corporate Capitalism* (New York: Alfred A. Knopf, 1977).

²⁵Dorothy Nelkin, *Technological Decisions and Democracy* (London: Sage Publications, 1977); and Guild Nichol, *Technology on Trial* (Paris: OECD, 1979).

9
Industry and Energy: Moral Dimensions of the Tasks
Alasdair MacIntyre

Morality has recently become fashionable in this country. Business and labor leaders make speeches in praise of it; ethics committees in legislatures exhibit unusual bursts of activity; courses in the ethics of this and that multiply in universities. We are in one of those phases, recurrent in American history, in which morality has been rediscovered yet once again. Such phases are perhaps even more dangerous to morality itself than those other periods in which only the most conventional and perfunctory homage is paid to it. For one thing it is characteristic of such phases that a hunt for scapegoats ensues, an attempt to identify individuals or organizations whom the rest of us can happily and indignantly blame: Richard Nixon, Richard Helms, the FBI agent who acted without a warrant, some corporate vice-president who gave or took a bribe. I hope that my use of these examples suggests immediately that my point is *not* that the victims of scapegoatery are usually innocent of that of which they are accused. On the contrary—very often, although not always, they are in fact guilty. My point is that scapegoatery is an activity which by directing our attention exclusively to the crimes and follies of particular individuals focuses our concerns in just the wrong way on precisely the wrong issues. Blaming and punishing individuals becomes a substitute for asking what it was in the structures of our common life which at the very least made possible and perhaps even positively engendered moral fault and failure; by fastening attention upon *them* we self-righteously distract attention from *ourselves*, from what it is in the forms of life in which we all participate which makes all of us sometimes morally impotent and some of us morally wicked.

Moreover the present fashionable concern with morality in general and the practice of scapegoating in particular is concentrated almost exclusively upon actions which are breaches of the kind of moral rule which tells us only what we ought *not* to do. Of course such negative rules *are* important, but reiterating them may not be a very good way to emphasize their importance. I do not believe that the reasons for the large moral insensitivities of the Watergate and White House conspirators or of the corporate bribe givers and takers arose from their not having been told often enough or forcibly enough that it is wrong to bribe, to steal and to corrupt. We all of us, or almost all of us, in fact know these things perfectly well; the problem is rather what it is that enables us to ignore or bypass this knowledge in crucial social situations. One reason may well be that our very concern with and emphasis upon negative, prohibiting rules leads us to lose sight of what is in fact centrally important to morality and thus in turn to fail to give their due importance to these selfsame negative, prohibiting rules. What is in fact centrally important to morality?

NEGATIVE AND POSITIVE ASPECTS OF MORALITY

Consider the kind of misunderstanding that would be involved if someone were to believe that the moral content and point of a marriage is to be discovered by asserting the rule prohibiting adultery and also to believe that *that* rule is nothing more than a special case of the rule which prohibits us from breaking our promises. Clearly what is wrong with adultery is not *primarily* that it is a breach of a rule; rather it is the case that adultery is an infliction of a gross injury upon a relationship of a certain complex kind, a relationship of caring, a relationship the making and perfecting of which is the activity which gives point and purpose to a marriage. It is because and *only* because of the particular goods and perfections which provide the institutions of marriage and the family with their point and purpose—goods which cannot be achieved outside those institutions—that the negative, prohibiting rules themselves have any point and purpose. If someone were to lose sight of this connection, if someone came to understand the morality of marriage exclusively in terms of a set of general, negative, prohibiting rules, then on being presented with some reason for breaking them, perhaps an attractive, delightful, flattering reason, he or she would be unlikely to see the point of *not* breaking the rules on this particular occasion; for he or she would somewhat earlier have lost sight of the point of keeping the rules at all.

I want to present a general thesis derived from this example. When the basic rules of morality in some particular institutional milieu—the home, politics, industry, the universities, or the professions—are not given the kind of respect which is their due, we ought always to inquire first if it is not because in that milieu men and women have become generally blind to the connection between the secondary negative, prohibiting rules of morality on the one hand and on the other hand the primary positive goods and perfections which it is the specific moral task of those inhabiting that kind of institution to achieve. For when we lose sight of the positive tasks of morality, then naturally enough the negative prohibiting rules will be more and more apt to have the appearance of mere arbitrary taboos. And when that happens to a large proportion of the inhabitants of a particular institutional milieu, it will be to a large degree a matter of the accidents of circumstance, upbringing and opportunity *which* individuals are flattered, delighted, or otherwise seduced into breaking the negative rules. But, if all this is so, it is easy to understand the irrelevance of scapegoating. The important questions are *not* concerned with *who* is to blame (I am not suggesting of course that those questions are completely unimportant) but with what it is about our common life that has led us to be generally blind to the place of the positive moral tasks which constitute our institutional responsibility—in the family, in the school, politics, in the place of work—and, for those in the electric power industry. Hence the irrelevance of the kind of business ethics which consists of little more than a reiterated praise of the negative prohibiting rules and which interprets such virtues as honesty and integrity as little more than systematic abstinence from rule breaking. If I am right, that kind of ethics is in fact part of the disease of which it professes to be the cure.

MORALITY AND THE ELECTRIC POWER INDUSTRY

But now, it will be asked, what does all or any of this have to do with the electric power industry? For the record of members of that industry in abstaining from breaking such negative, prohibiting rules is by and large an exemplary one. Indeed I should find it entirely comprehensible if some members of the industry with the negative conception of morality which I have just described should regard it as gross impertinence for a professor of philosophy—or anything else—to come and lecture *them* on business or industrial ethics. "We are looking after our own moral concerns perfectly well," they might say with some justice. "We do not come into universities to give lectures on *academic* ethics. Why should you dare to give lectures on *business* ethics to us?" The point is well taken. It would neither worry nor surprise me if it were suggested that standards of morality are probably a good deal higher among the officials and servants of public utility companies in general and among the executives of the electric power industry in particular than they are among professors, and I should be very happy to see *them* coming into universities to give lectures on ethics to *us*. But I hope that what I have already said will make it clear than I am not inclined to blame, to scapegoat, or indeed to moralize. Paradoxically enough, insofar as I am inclined to question the moral resources of the electric power industry, it is because in one way its moral record is as good as it is. For the morality of the industry has been essentially a negative one, one of proved abstinence from wrongdoing.

The tradition of the electric power industry since the end of the Great Depression has been very much at one with that of public utilities generally. It gave a rather rigid interpretation to its legally mandated task, at the heart of which was the requirement to supply electric power on demand. This requirement led in turn to a need to anticipate demand by making accurate predictions; what was not perceived often enough was that the activities which generate such predictions are themselves one more factor in creating the climate which produces universal demand for electric power. Thus the American electric power industry not only supplied power with a success unparalleled in human history, but also participated in creating the demand that made that success necessary to the rest of American industry. All this was achieved in such a way that neither the more general question, "Growth for what?" nor the more specific question, "Electric power for what?" needed to be raised, let alone answered. Those questions implicitly were held to be questions only for consumers, just as questions of the legal constraints to be imposed on the provision and use of electric power were held to be questions for Congress and for citizens. Of course those who worked in the electric power industry were themselves also consumers and citizens; but one outcome of this kind of understanding of the tasks of the industry was a self-imposed distinction between their role in the industry and their roles as citizens and consumers.

What then was the morality required and practiced by the industry? It was generally a strict morality of nonintervention in every area but that which it and others regarded as its own legitimate realm. It was a morality of faithful compliance with the law and the will of Congress on the one hand and with contracts and the will of the consumer on the other. Morality was very clearly envisaged as providing a set of constraints within which the executive set about his industrial and commercial tasks. There was one positive aspect to the morality—its basic assumption, so firmly held as scarcely needing to be stated, was that the providing of electric power within the limits set by these negative constraints was an unqualified good. Environmental concerns did not emerge until the very end of the period about which I am speaking, and the obvious connections between the provision of electric power and the possibilities both of providing employment for an increasing workforce and of increasing the comfort of domestic life reinforced this assumption. The consequence was that a whole range of decisions came to be treated as purely or almost purely technical decisions to be handed over to the economists and the engineers: What types of plants should be built? Where should they be sited? At what points should investment be made? What skills did the industry need? Congress provided the mandate, the consumer provided the data for prediction, the executives of the industry provided the commercial integrity and the skills for answering such questions, but the questions themselves were technical, not moral.

The time has come to make clear the relevance of the first section of my argument about the nature of morality to the second section of my argument about the moral record of the electric power industry. In the first section of my argument I suggested that morality has at least two parts: a negative part which consists of prohibiting and constraining rules and a positive part which consists of setting before us goods and perfections which will provide our tasks and projects with point and purpose. What the second section of my argument suggests is that the electric power industry—and in this it has been no different from the rest of us in modern society—has allowed its moral perspectives to be defined too much in terms of the negative prohibiting and constraining rules and not enough—and not explicitly enough and not reflectively enough—in terms of the positive goods which ought to inform its tasks. It is those goods which provide the distinctive moral dimension in any definition of the future tasks of the electric power industry. We can very happily take for granted the need to observe the requirements of the negative rules; what we do need is a more explicit assertion of the industry's positive moral tasks.

Yet before turning to a statement of these tasks, it is worth emphasizing briefly one further consequence of the first section of my argument, that the search for scapegoats is always a symptom of a defective mode of moral analysis. It scarcely needs remarking that the discussion of energy policies in the last few years, not excluding the discussion of electric power, has been distinguished in some quarters by an almost obsessive search for scapegoats. Nor has this search been the monopoly of any one side in the argument; there has been enough opportunity for scapegoating for everyone to take part. OPEC, Saudi Arabia, the oil companies, the federal bureaucracy, and environmentalists have all been convenient whipping boys. My argument does not of course imply for a moment that any one of these is without fault and error. Transnational corporations, modernizing governments, and federal civil servants do indeed all have their own distinctive ways of doing harm in the world, as do professors. Nonetheless, scapegoating leads us to precisely the wrong sort of analyses, just as it arouses the most useless kinds of emotions. The argument has to begin somewhere quite different. To what goals, stated positively, ought we all to be contributing? And what is the moral dimension of the pursuit of those goals? To the answer to these questions I therefore now turn.

THE INDUSTRY'S POSITIVE MORAL TASKS

One positive moral task of the industry is to assume a large public responsibility in areas which it has hitherto treated with a scrupulous but, if I am right, partly misplaced, respect as the responsibility either of the Congress and the executive branch of government or of the industrial commercial and private consumer. That responsibility is to urge, cajole, and compel our society to make certain choices and to make those choices in as open, as explicit, and as rational a way as possible. There are two different kinds of reasons why this responsibility falls to the electric power industry. The first is concerned with its unrivaled strategic position in the supply of power and with the scale and scope of its resources. Those resources are not merely or most importantly even a matter of plant and facilities; what matters even more is the industry's accumulation of the relevant kind of knowledge and relevant kinds of skills. The choices in the matter of energy policy which our society now has to make are going to be imprecise and ineffective if they are formulated at the level of the kinds of rhetorical generalities to which we are too often treated; those choices, if they are to be clear and effective, must be between alternatives that have been presented in specific and detailed terms. Both government and the general public have to learn what the full range of alternatives in the use of each particular type of source of energy is under the relevant specific local conditions of particular communities and environments and the possible consequences of each type of use of such sources. Engineering, economic, environmental, and social considerations are going to have to be presented within a common framework of political choice. And both government and the public have to be educated into a realization of what the choice of each alternative excludes from our common future as well as what it includes.

It is perhaps obvious that the members of the electric power industry are as a group uniquely well fitted to present such choices. It would of course require of them that they transcend their position as one special interest group among others. And it is a task that could only be carried out effectively with the participation of the entire industry, of labor as well as of management, of those who engage in research and those who supply power, of those who work at national and regional levels and of those who work at the level of states, cities, and small local communities, each of whom would have to transcend his position of special interest within the industry in order to participate in a dramatic and effective educational dialogue with the whole society. But, it may at once be asked, how can we expect those who work in the electric power industry to transcend the limitations of their own special interests? Are we not asking them to become moral supermen? There is no simple, glib answer. But part of the complex answer is that we have asked no less from a variety of special interest groups in times of national crisis and especially during World War II. There is no doubt that the present crisis may present at least as intense and certainly a more extended threat to our national identity and our commitment to those shared values which make a common life possible than even that war. Another part of the answer is that a necessary and, I hope, not impossible condition will be to bring together the ways of thinking—as power company executive, as consumer, and as citizen—which the patterns of corporate life have assigned to distinct and overcompartmentalized roles. This is clearly something that cannot be brought about overnight by some heroic act of the moral will. Rather it is the case that if the electric power industry were to set itself the moral task of educating our society in how to make certain types of important choices, it would coincidentally find that it had also assumed the task of reeducating itself. It is indeed only through transforming the external world of society and nature that we also transform ourselves.

I have argued so far that the electric power industry has a special responsibility to initiate and to forward a dialogue on how to make choices with respect to energy because of its own resources and strategic position; but there is a second type of reason for suggesting that the industry has this special and relatively new responsibility. It is that those so far who have been conventionally assigned this task by the democratic process have signally failed to discharge it in any but the most inadequate ways. Our political institutions and the media have so far functioned very badly as agencies for the formulation and making of public choices in this area. It is not difficult to understand why.

THE WEAKNESSES OF OUR DECISION-MAKING INSTITUTIONS

We have in our society by and large two and only two institutionalized methods for coordinating individual preferences and transforming them into public choices, those of the market and its allied institutions on the one hand and those of government legislation, taxation, and regulation on the other. It is scarcely surprising that public debate has focused explicitly and implicitly, in energy questions as elsewhere, on the choice *between* those two and that the form of public debate has been largely that of an indictment of each of these methods in turn by the proponents of the other. The sad fact of course is that both parties are right. Both methods are grossly defective and specifically with respect to the kind of choices which our society now needs to formulate. For where both most obviously fail is in providing for the presentation to individuals of the range of alternatives which confront them and between which they have to choose, under the constraints of time, available resources and human skill and ingenuity. Consider first some defects of the mechanisms of the market.

Two virtues commonly ascribed to a free market are first that through its mechanisms producers respond to the free choices of consumers and secondly that resources are by means of this response efficiently allocated. It is of course in response to changing *prices* that both these effects are produced—insofar as they *are* produced. But now consider the key phenomenon of investment. By investing now rather than spending we exercise a choice in favor of future goods rather than present goods. Thus, as Kenneth J. Arrow has recently written, "goods to be produced in the future are effectively economic commodities today. For efficient resource allocation the prices of future goods should be known today. But they are not." And hence, he adds, "plans made by different agents may be based on inconsistent assumptions about the future."[1] Arrow's argument is explicitly directed against the view that markets are efficient allocators of resources, but his premises equally warrant the conclusion that the sovereignty of the intended consumer is to some large degree illusory. For what will determine future movement in prices will be in key part the mistaken beliefs of investors about future prices which *ex hypothesi* are not responsive to the real choices of consumers.

A crucial premise in Arrow's argument is of course that the prices of future goods are not—and to a large degree cannot be—known today. All our data about economic forecasting in general confirm the truth of this premise. Economic prediction and social prediction in general have a bad record. No economist predicted stagflation before it occurred. Monetary theorists have failed to predict rates of inflation correctly. And it has been shown that the forecasts produced for OECD since 1967 on the basis of the most sophisticated economic theory available were less accurate than would have been arrived at by naive, commonsense methods of forecasting.[2] I cite this evidence *not*, let me emphasize, to cast any particular scorn on economic forecasting; among types of social forecasting it stands out for the ability of its predictors to learn from their past mistakes. Nonetheless it provides no evidence and, so far as can be seen at present, has no likelihood of being able to provide in any foreseeable future evidence to undermine Arrow's premise.

Markets then do not provide anything like adequate mechanisms for coordinating individual preferences where we are concerned, not with preferences for the present or for a future so immediate that we may think of it as almost present, but instead with a yet-to-be-shaped future. Equally they only provide mechanisms for coordinating individual preferences, once those preferences have been formulated and expressed in acts of consumption. The market therefore is of no help to us in those areas of life where we have to decide what our patterns of consumption are to be, how our preferences are to be ranked, how our desires are to be ordered. It is *within* families and households, schools and hospitals, corporations and labor unions, not in the market, that such decisions have to be made and characteristically are made.

These two weaknesses of the market are both of obvious relevance to the questions of how the debate on energy is to be conducted and who is to conduct it. For the debate on energy is peculiarly concerned with the future—it is in fact a debate centrally about investment; and it is a debate that has to be conducted within what Daniel Bell has felicitously called "the public household." We need to reason together in order to discover with what choices we want to enter the market. We are, that is to say, at a point in the argument where to tell us to rely on the mechanisms of the market is not so much mistaken as irrelevant.

Those who believe in the imposition of some political solution through legislation and regulation are likely to agree all too readily with this diagnosis of the weaknesses of market mechanisms as decision procedures. But it is important to see that there are weaknesses and defects in our system of political decision making which correspond precisely to the weaknesses and defects of the market. It, like the market, is responsive to the pressures of the present much more rationally than to those of the future, partly because, as I noticed earlier, our ability to predict the future accurately is in general extraordinarily limited. And it, like the market, once again is far more effective at expressing already formulated choices on familiar issues than at formulating new possibilities of choice in unfamiliar areas. Both the market and the political system have a vested interest in the present and the spectacle of continuous technological innovation with which our culture confronts us tends to conceal from us the extraordinary conservatism of even those patterns of innovation.

To have pointed to the weaknesses of the market and of our political system is not of course to suggest that there is any alternative to them. Like Mount Everest, they are simply *there*, inescapable features of our social landscape, and in any case it is perhaps true to say of them so far what Winston Churchill said of democracy in general, that it is the worst form of government, except, of course, for all the others. But nonetheless what is quite clear is that neither the market nor the political system will provide the ordinary citizen with an adequate arena for formulating and expressing radically new choices of the kind that the energy debate thrusts upon us. And if public discussion which will enable ordinary citizens to formulate and express radically new choices does not take place then the political and economic outcomes will inevitably be sadly defective. But who is to set on foot the debate necessary to supplement our conventional economic and political institutions? Who is to educate the educators? I see very few individuals or institutions who are both capable of taking up this task and who possess the resources and the strategic position to carry it through apart from those individuals and institutions who comprise the electric power industry. Their work compels them to interact with both producers and consumers at essential points. They have a peculiar responsibility which arises from the fact that if they do not discharge it, it is unlikely that anyone else will. If the industry does embark on this task it will perhaps be accused of trying to preempt the democratic process; but if it does not take action that will render it liable to such accusations, the democratic process itself may fail us.

At this point another question might be interjected. How would our shareholders react to this responsibility you are urging upon us? There is no conflict. Because the energy problem is, as we shall see, one of investment, the greatest need of shareholders is to be provided with resources enabling them to make educated choices. The greatest need of the industry is to have educated shareholders to whom it can be responsive. The industry's responsibility to the general public points it in the same direction as its responsibility to its shareholders.

WHAT MORAL VIRTUES DO WE NEED TO CULTIVATE?

What, then, is the task and why does it have moral dimensions? In order to answer this question I shall have to state briefly what I take the energy crisis to be about, which is certainly not a shortage in absolute terms but pressures on the prices to which we have become accustomed. If we must shift investment toward new energy sources something else of importance to our lives will lose out, which is why the energy crisis involves our whole way of life. What OPEC is saying to us is in essence what the Spaniards, in one of their proverbs, assert that God has said to them, "Take what you want, and pay for it." But in changing investment patterns, choices must be made between competing goods. How should we weigh *local* control and autonomy over *national* needs? How important is it to provide full employment? How will the pricing of farm products affect the family farm and is it important? Ought we to allow defense commitments to become extended through reliance on foreign energy sources? What are our responsibilities to the next generation and to the tenth generation after that? What are our responsibilities in Asia and Africa? How ought we to decide questions about responsibility? Every one of these disparate questions which our conventional habits of mind dangerously often lead us to think about in quite different ways and contexts as though the answers to each were independent of the answers to all the others are questions the answers to which have both immediate and long-term implications for the ways in which we produce and use electric power. Each of these questions is therefore one in the formulation of which and the assessing of the range of possible answers to which the voice of the electric power industry needs to be heard. I do not mean by this of course that I think that the industry—or anyone else—is capable of instant expertise on Africa and future generations and defense policy; I mean rather that it has to be brought home to us that questions about all of these are also questions about energy use in general and about electric power in particular, and it is on this aspect of all these questions that we need to be educated into new habits of mind.

It is not difficult to see that when we rethink this wide range of questions as questions about energy, a number of different kinds of moral concern must arise. The first is a concern for *complexity*. Complexity is not normally thought of as a notion with moral implications. But oversimplification, the sacrifice of complexity, is in fact a crucial form of the vice of untruthfulness. Yet this is not the only morally damaging harm that is likely to arise from oversimplification of the issues. Just because so many different kinds of issues of policy and practice interlock at the point of energy use, there is no simple way to assess the costs and benefits which will arise from any particular proposal. How, for example, are we to weigh as considerations relevant to the same proposal the harm of damage to the environment against the harm of making fewer jobs available to those who badly need work and both of these against the harm of injury and death to a certain number of presently unidentifiable individuals? Our culture possesses no rational means of weighing such evils one against the other. Because it is a secularized, pluralist society which prides itself on maintaining the freedom of every contesting moral voice to participate in public debate, as a society we possess no adequate, shared, rationally defensible set of standards by which to judge goods and evils. Hence it is always much easier for us and more tempting to us to consider these issues in a piecemeal, local way in which some particular compromise—determined through bargaining by local circumstance and balance of power—will in the short run at least satisfy the particular, local contending parties. Our whole political and legislative process is biased in favor of such bargaining procedures and outcomes; that is to say our whole political and legislative process is biased toward oversimplified statements of problems (oversimplified because they are not often enough formulated with an adequate recognition of the large-scale general issues which confront the whole society in the long run) and consequently not only toward oversimplified solutions, but unjust ones. The lack of a sense of complexity leads not only to untruthfulness, but also to injustice. Why is this so?

I asserted earlier that our culture possesses no general set of standards which will enable us to evaluate costs and benefits of very different types in a single rational argument. This makes it all the more important that our evaluations do satisfy two minimal requirements of justice. The first is that everyone relevantly involved—and where energy is concerned that means everyone—should have a chance to say what is to count as a cost and what as a benefit. The second is that, so far as possible, those who receive the benefits should also be those who pay the costs and vice versa. Neither the former nor the latter principle has received anything like adequate attention in recent debates. When environmentalists urge policies which will as a matter of fact significantly reduce the number of jobs that might otherwise be available, it is rarely, perhaps never, the case that they are able to show *either* that those who would lose their opportunities of work are the same people who would benefit from the environment *or* that those who would lose their opportunities to work have had a part in shaping the conception and criteria of costs and benefits involved. Environmentalists in the present have sometimes been as untrammeled in endangering other people's jobs as industrialists in the past were in endangering other people's environments. Equally, when opponents of nuclear power stations urge policies which will as a matter of fact significantly increase our future dependence on coal, the lives that will be lost as a result of *their* policies are not their own.[3]

The power of the British National Union of Mineworkers in a nationalized industry has been used as effectively as is humanly possible to lower the risks of mining deaths, so that in the following example, any British comparison understates rather than overstates the balance of risks between coal mining and nuclear power use. John Lyons, general secretary of the British Electric Power Engineers Association, summarized the relevant part of the report as follows:

> We calculate that taking account of the mining, transport, and emission risks for both fuels, the production of 55GWe of electricity from nuclear power was overall seven times safer than the same output from coal field stations. In human terms this means that for each year that 55 GWe of electricity is generated by coal field plant about 350 eventual deaths could result amongst miners and the general public, whereas using nuclear generation this figure could be 50 at most.[4] Thus justice demands not only that everyone has a voice and a vote, but that some people—the populations that will bear the relevant risks and dangers, or most of them—have more of a voice and more of a vote than others.

There is at least one further requirement that must be met if justice is to be served. I have already identified the energy crisis as primarily a crisis about the investment of resources and I have now urged that we try as far as possible to distribute costs and benefits in a way that not only is just, but accords with the sense of justice of our population. Yet ordinary citizens are going to have no opportunity of understanding how the costs and benefits of different investment policies do and will impinge upon them, if the costs are systematically concealed by subsidy. For most Americans—as for the inhabitants of most advanced countries—the cheapness of energy to the consumer in the past fifty years has disguised its true costs. From this point of view the price rises imposed by OPEC have been very largely beneficial, and the masking of the true costs of energy for the population by government subsidy and manipulation is highly dangerous. Ordinary citizens respond to their everyday experience and if their everyday experience conceals what they are really paying for energy in all sorts of disguised ways, then no amount of theoretical education is likely to succeed. Of course we cannot allow the ordinary citizen's life to be ruined by exposing him or her to rises in energy costs which impose an impossible burden; but it would be far better to assist poorer citizens by tax reliefs—even including a negative income tax—than by the kind of subsidy which distorts both prices and our understanding of them. We can only decide both democratically and rationally how it would be just to distribute costs and benefits if there is a widespread perception of what the costs and benefits of different energy policies are.

So far I have been specifying some of the requirements that must be met if the virtues of justice and truthfulness are to receive their due. But these are not the only virtues that we need to make central to our tasks in relation to energy production and use. And some of the virtues that we need to cultivate do not appear on traditional lists of the virtues. One is the ability to live with unpredictability. I have been arguing the need to consider our energy problem from the standpoint of the whole society rather than from particular and local standpoints, and I have noticed that our resources for social and economic prediction are very meager. Hence what must seem initially a paradox: we need a national—indeed an international—vision and policy and thus we need to plan on a large scale, but we also have to recognize that our plans are for a future which is always apt to surprise us with its unexpectedness. This unexpectedness has a number of roots of which it is necessary to notice two here. In part our social future is unpredictable because of the ways in which the future development of mathematics, science, and technology are unpredictable. Sir Karl Popper and others have produced important arguments to show that a certain kind of radical conceptual innovation cannot be predicted.[5] An obvious and obviously crucial example is the mathematics of Turing or of von Neumann, work essential to the development of modern computing science and engineering whose outcome could not have been rationally predicted in advance. In a science-based society such as our own the unpredictability of fundamental scientific discovery clearly generates one kind of social unpredictability. And there are other kinds.

In changing the social world and its national environment we also change ourselves. No matter how sophisticated our social planning, with respect to energy or to anything else, we can never rule out the possibility that in the course of implementing our plans we shall acquire views, interests, and desires markedly different from those which led us to draw up and implement our original plans. Hence it is crucial that our planning not be so inflexible that it leaves our future selves—let alone succeeding generations—with no or few options acceptable to them because we were at an early stage too rigidly insistent on what is now acceptable to us. Burke reminded us that we *cannot*, even if we wish to, disenfranchise the dead; we have to remember as urgently the importance of enfranchising the unborn, of allowing the future too to have a voice and a vote. It follows that there are great moral as well as practical dangers in making too many large-scale irreversible investment decisions which foreclose on future choices. This is especially a danger because the urgency of our immediate needs always tempts us toward short-term solutions. Hence the importance of the debate that followed Amory B. Lovins' article in *Foreign Affairs* for October 1976 on "Energy Strategy: The Road Not Taken?" and the other works in which Mr. Lovins has urged upon us what he calls soft rather than hard energy policies.

Let me pause here to notice one feature of the examples that I have used so far to illustrate the moral importance of certain types of consideration, of certain types of virtue. These examples have been naturally enough drawn from recent episodes in the energy debate and from opposing sides. Hence some of what I had to say about justice might seem supportive to the advocates of massive investment in nuclear power, while what I am now saying about the need for an ability to live with unpredictability may seem to endorse what some of their critics have said. But what matters at this juncture is not so much the particular policy implications of particular points; rather it is the need to underline the fact that we ought not to expect moral considerations all to point in one single policy direction; there is a genuinely tragic dimension to the energy debate because any particular policy direction is going to involve the sacrifice of some authentic goods for the sake of others. To have noticed this is to be able to restate in another way a point that I made earlier in the argument when I asserted that we possess as a society no rational way to make large-scale complex decisions affecting our whole social order, just because we lack the relevant kind of shared moral standard and shared moral perspective. It is now clear that to say this is also to say that our culture as a whole lacks any clear sense of how to handle tragic situations, situations which reveal our moral and human limitations in relation to the tasks imposed upon us.

THE WHOLE SOCIETY AS THE FOCUS OF THE INDUSTRY'S RESPONSIBILITY

It is, I believe, because we have never learned to face up to our moral limitations that we have lost to such a large degree in this society our vision of positive possibility. We have trudged for so long into a future of apparently limitless consumption that we do not now find it easy to remember who we are and what links us to others. For the notion of possibility is always the notion of some future form of community which provides us now with standards and goals by which to diagnose our various forms of inadequacy and to set about remedying them. Just such a common vision—partial, not always coherent, but providing the essential sustenance for our constitution—was implicit in the founding of this republic. It is there to be read in Jefferson's letters and papers in one version, and in *The Federalist Papers* in another. And just such a common vision was already present, perhaps even more strikingly, in the first foundation of the Massachusetts Bay Colony. We today lack—and the result is an emptiness at the heart of our political and cultural life—what seventeenth-century puritans and eighteenth-century heirs of both Christianity and the Enlightenment still possessed.

What I suggest is to press forward in the public arena the debate about energy in such a way as to make its moral dimensions clear so far as possible to the whole society. This will reveal to our society that it is, to a degree which a good deal of political rhetoric conceals from us, involved in a moral crisis as well as in an energy crisis. We do indeed have great moral resources. Americans are able to face the truth about themselves in a way that has not always been true of other peoples; they do not need to be flattered by being told how good they are. Nonetheless the prospect is a dismaying one and requires courage. We do not know how to reason together morally in an effective way. And this lack—just because it is something wanting in the social order as a whole—will never be remedied unless we face it as a society.

If all this is true, then I am inviting the electric power industry to accept what is both a great and yet in some ways an inescapable responsibility. For it cannot adequately formulate, let alone resolve, its own problems without confronting the moral dimensions of those problems, and it cannot confront the moral dimensions of those problems without laying bare the dilemmas and defects of the whole society. In doing so every consideration that I have argued as morally relevant for the industry turns out to be as relevant for the society as a whole: the need to understand morality primarily in positive and only secondarily in negative terms, the damages of resorting to moral scapegoating with its concomitant easy resentment and instant indignation, the need to extend the discussion in a way that neither the mechanisms of the market nor our conventional political institutions serve easily or adequately, the importance of constructing a new table of the virtues in which the relationship of a sense of complexity to justice and the relationship of an ability to live with unpredictability to responsibility to and for the future find their due place, and above all the need for an integrative vision of a possible form of future community which will at once provide a *telos*, a goal toward which our actions and policies may move us and which will also furnish us therefore with standards by which to measure those actions and policies in ways not now open to us.

For any human group to have been placed at such a strategic point in American history and indeed in human history, as the electric power industry has been placed through no will of its own, will certainly often seem something of a burden in the coming years. If the opportunity is to be used creatively it will mean, as I noticed earlier, undertaking to break with established, compartmentalizing habits of thought both in our academic disciplines and in our political and social life. It will involve conflict within and outside the industry. It will mean thinking about the relation of the role of citizen to the role of industrial executive to the role of consumer in quite new ways. When the assumption of such a burden of innovative self-transformation seems far too much to ask, it will be important also to remember how great the cost of *not* assuming that burden will be. The industry cannot continue as it has done, discharging in a quiet and orderly way its allotted tasks within self-prescribed limits under narrowly conceived legal and economic constraints. Failure to move toward the larger role will inevitably mean that the industry will come under growing pressures from without which it cannot control and which may be exerted without any control whatever. The tasks which morality sets us are always difficult and sometimes, as Philip Morrison reminded us, dangerous. It is a genuine if small consolation to remember that the cost of not assuming them is always a loss of self-mastery and a surrender of self and community to impersonal, destructive fates. If we fail to devise adequate energy policies we will lose our autonomy to the wrong kind of bureaucracies, to unstable foreign powers, and to a less than rationally molded public opinion—in fact, to the politics of panic. Industry, like the rest of us, has something to fear and much to hope. To the tasks of creative scientific and technical thinking it must now add the tasks of moral intelligence.

NOTES

[1] Kenneth J. Arrow, "A Cautious Case for Socialism," *Dissent* 25 (Fall, 1978): 248.
[2] Alasdair MacIntyre, *After Virtue* (London: Duckworth, forthcoming).
[3] Electric Power Engineer's Association (United Kingdom), *Some Implications of Radioactive Waste from Nuclear Power Generation in the UK up to the Year 2000*, 2 vols. (1978).
[4] *The Times* (London), November 29, 1978.
[5] MacIntyre, *After Virtue*.

10
Science, Technology, and Social Achievement
Philip Handler

Although it took professional economists overlong to appreciate that technology has been the principal engine of social change and economic development in the industrialized world, that notion now appears to be well established. Indeed, only economists now seem to express the optimism that many of us once shared—belief that science-based technology would continue to drive change, that economic growth would continue and concomitantly bring with it betterment of the human condition. As Harvey Brooks has noted elsewhere, faith in that axiom is declining in the scientific and technical community. Some of us cling to the belief that things will work out well, that human ingenuity will produce new technologies which will alleviate the problems associated with the old and provide new and unanticipated benefits. Others consider that the unanticipated consequences of technology have already injected the seeds of inevitable disaster into human affairs. I seem to find myself in a midposition. While I hold guardedly to the optimism to which I was conditioned by long experience, it is now a rather conscious preference and almost deliberate rejection of qualms and doubts which assert themselves.

There is no need to rehearse the variety of technologies that have molded the societies of industrialized nations in the last few centuries, with social consequences nowhere more dramatic than in the United States. I shall treat this process in terms of an individual dimension and an institutional or distinctly social dimension that is somewhat different but which clearly intersects the first. While the distinction is rather artificial it highlights the fact that along the individual dimension the accomplishments of industrialized nations, the powerful impact of science and technology has been extraordinarily beneficial: an immense transformation of the quality of individual life made possible by the availability of electric power, the internal combustion engine, and a thousand other technologies. But in their institutional dimension, the contributions of science and technology seem much smaller and, in some ways, possibly negative.

INDIVIDUAL AND INSTITUTIONAL ASPECTS OF ACCOMPLISHMENT

The distinction I wish to rely upon is rather simple. Human progress refers to improvements in the circumstances of individual members of society—their health, the conditions under which they live and work, and their capacity to realize their individual physical and intellectual capabilities. Institutional progress occurs through improvements in the functioning of social institutions, enabling members of a society to interact more effectively with each other as well as with members of other societies, for example. There are, of course, tensions and trade-offs between the two; when nations go to war, individuals pay the price.

Intrinsic Scientific Factors in Individual Progress

The sheer joy of scientific knowledge has been made available to a significant fraction of humanity. We may complain about the extent of scientific illiteracy and about cultural gaps separating those educated in science from those who studied the liberal arts. Nonetheless, the lives of hundreds of millions of persons, worldwide, have been immeasurably enriched by some knowledge of science—of DNA, of atomic nuclei, of the moon and Mars and Jupiter, and even of "black holes," of what we are and where we are and how we have constructed the man-made world. The sense of understanding, faith that what is not understood today will be learned tomorrow, is a vastly different life experience than is the ignorance that engendered mysticism, superstition, and black terror.

The dramatic improvement in the health of individuals in the last century, surely one of the proudest accomplishments of our civilization, was achieved in the main by relatively unsophisticated science and technology. Managed supplies of pathogen-free water, plumbing in less crowded households, and improved nutrition were the primary contributors to the remarkable decline in age-adjusted death rates—from 33 to 8.9 per thousand persons in the century 1870 to 1970. But improved sanitation followed upon knowledge of bacterial and viral pathogens and the chemical and engineering science required; improved housing followed upon increased family income, education, employment, transportation, electric power, and fabricated materials. As late as 1940, in the less advantaged fraction of the American population, 80 percent of black households and 40 percent of white households lacked some or all plumbing facilities. By 1974, these figures had declined to less than 10 percent of black, and only 2 percent of white households.

Agricultural and Industrial Productivity

Improvement of nutritional status reflected scientific knowledge of the consequences of malnutrition and sufficient income to purchase an adequate balanced diet, but also an agribusiness enterprise that provides an abundance of diverse foodstuffs at relatively low cost and assures their nationwide distribution, which in turn depends upon water supplies, fertilizer, pesticides, irrigation, tractors, and a fleet of refrigerated trucks and railroad cars. The incidence of frank nutritional deficiency in the industrial nations has declined almost to zero. To appreciate that fact you must realize that when I moved to North Carolina in 1939, pellagra was the leading cause of death in eight southeastern states and every year filled every bed in every mental institution in those states. In the United States today, cholera and typhoid are almost unknown, malaria and yellow fever have not been seen for more than a half century, and the incidence of infantile diarrhea is steadily declining. While the primary contributor to the improvement in mortality statistics is the decline in the infant mortality rate—from 50 per 1,000 in 1937 to 15 per 1,000 in 1977—nevertheless, health in the middle years has so improved that between childhood and 45 accidents are the most prominent taker of lives. Illness is not unknown, but a combination of drugs and surgical skill has made the middle years of life relatively uneventful and free of pain for most of us.

Increased productivity on the farm, the result of a family of technologies, made possible the industrial labor force. No statistic concerning the United States seems more dramatic than the fact that when our national population was 100 million, the farm labor force was 14 million; after the population had doubled, the farm labor force had declined to 4 million. And life on the farm now enjoys virtually all the domestic amenities available in the city.

Industrial productivity, increasing at about 3.5 percent per year over many decades, induced the abolition of sweatshops and a decline in the number of working days per year and working hours per day, while increasing family income sufficiently to allow our young opportunity for increased schooling. Only 30 percent of pupils finished high school in 1924 as compared with more than 75 percent by 1966. Among persons aged 25 to 34 years old, the proportion of whites who have completed four years of college increased from 10 percent to 20 percent in the decade 1964 to 1974; the rate for blacks also doubled in that period but another decade will be required to get the proportion of black college graduates up to 15 percent. With increased productivity came increased income.

Median family income, about $7,000 per year for white families in 1947, rose to over $16,000 by 1978. Moreover, the largest component of this change was the narrowing of the gap in income between the lower and higher paid members of society. The number of individuals possessing gross wealth of more than $60,000 increased from 2 million in 1953 to 13 million in 1972.

Decreased working hours and increased income engendered leisure time activities which have become a significant component of the economy. The number of books published annually in the United States increased from 11,000 in 1950 to 40,000 in 1974. Someome must buy and read them. Fourteen percent of American adults cite reading as their favorite leisure activity. The number of symphony orchestras increased fivefold in the two decades following 1955; statistics concerning sporting, picnicking, and fishing equipment, outboard motors, boats, foreign travel, and theatergoing are staggering compared with any prewar experience. Our combination of wealth and leisure time has wonderfully enriched the lives and experience of a large fraction of the population. Television watching has become the favorite leisure activity of half of our people. Whatever the shortcomings of commercial television, it has expanded the personal experience of individual human beings in a manner unimaginable in previous history—and in so doing affected for all time the workings of history. Consider only the repercussions of the fact that a half billion pairs of eyes watched President Sadat as he entered Jerusalem a year ago or the consequences of the fact that the war in Vietnam was played out in our living rooms every evening.

Institutional and Social Progress

When we turn to indicators of institutional social progress, matters look much less rosy. Let me try a few indicators for the period 1960 to 1975, an interval during which the total body of scientific understanding approximately doubled. Changes in the status of women and of ethnic and racial minorities in our society in the last few decades need no recounting. We would certainly count this a proud achievement of American civilization. It may offend some of the activists who have led these movements but it is not unfair to indicate that the success of these movements was made possible primarily by technology and an expanding economy. To put it cruelly, our society was not kind to horses until we lost the need for live horsepower. Slaves were freed as a result of a vast national upheaval but it happened when slavery as an institution was altering from an economic asset to a liability. Legislation protecting children and women at work was passed only when the economy could tolerate it. The release of women from the thralldom of the kitchen finds its origin in the invention of the tin can as a means of transporting food for Napoleon's armies, culminating in the cornucopia of today's supermarket.

Women and the Family

It was the elimination of the need for human muscle power in the workplace, which we owe to technology, that opened industrial employment to women. The increase in productivity generated more jobs for them. Here is one institutional accomplishment with which we may be pleased. At the same time, we record growing instability in the family. The number of divorces granted annually in the United States increased by a factor of two and one-half between 1960 and 1975. In 1975 alone the lives of 1.2 million children under the age of 18 were exposed to the adverse effects of broken homes. Science and technology can be linked to this situation only indirectly. Still, more than one in every six children under 18 now lives in a single-parent family, including half of all black children.

Social Problems on the Margin

The extent to which unemployment derives from economic as opposed to structural, social causes may be debated. As a social problem, it is particularly serious for young black males, for whom the level is 25 percent, the highest level of overall unemployment during the Great Depression of the 1930s. But a considerable fraction of current unemployment has become structural. Belief is waning in the classical philosophy that semi-trained or untrained workers displaced by machines or automated processes will, statistically, find other employment as the economy expands. It is no longer neo-Luddite to consider such unemployment a permanent feature of the economy. In principle, it is remediable by education which would fit all persons desiring employment to find a place in a more skilled employment sector. But even that may not suffice. Since 1967, the proportion of all unemployed persons who indicate as a reason for their status that they simply "could not find a job" has steadily increased while the fraction citing "lack of training, education, or other personal handicap" as the primary reason has steadily declined to less than ten percent of all unemployed persons.

The rising personal aspirations that our society has enjoyed and our history of upward social mobility catalyzed by the public education system are now inversely mirrored in frustration and discontent. This is nowhere better revealed than in the statistics concerning crime. The homicide rate has more than doubled since 1960, reported robberies have nearly quadrupled, and aggravated assault more than doubled. Property crime rates have tripled and our homicide rate is nearly four times that of the country with the next highest rate in the industrial world (Finland). The annual production and import of small firearms for personal use has increased from about two million in 1960 to almost seven million in 1974 when these were used to commit 14,000 murders as compared with less than 4,000 in 1961.

The Situation of Institutions

As the society becomes more complex, and that results largely from technological advance, the established institutions of the society have come to be viewed as less competent in fulfilling their responsibilities. This coincided with such traumatic events as the war in Vietnam and Cambodia, industrial bribery and Watergate, and episodes such as Torrey Canyon and Three Mile Island. Respect for government, industry, educational institutions, the church, and the courts has steadily declined, along with confidence in their leadership. No longer are they reliably the pillars of society.

However loose the tie, certainly recent scientific and technological achievements have not contributed to the *more* effective functioning of our social institutions, our ability individually and collectively to relate to one another. Yet the period in question was the golden era of American science and only slightly less so for American technology. The very conditions under which science and technology have flourished are the very conditions that have posed major challenges to our social institutions and the fabric of social relationships on which they are based. The growth in the size of organizations and the resulting increased specialization of functions and roles necessary to support rapid scientific and technological development have created dramatically different and more difficult management problems at all levels of society.

Mass communication and the rising level of education of the population which result from, and contribute to, science and technology have generated greatly increased expectations, some of which cannot be fulfilled. Poverty is still *a* problem, but for the nation as a whole it is not *the* problem. This may be the explanation for the finding in surveys conducted by the Survey Research Center at Ann Arbor that "during the period between 1957 and 1972 when most of the economic indicators were moving upward, the proportion of the population who described themselves as 'happy' declined steadily and this decline was most apparent among the most affluent part of the population." Unfulfilled expectations in lower income groups and the sense of social malaise in upper income groups place great pressure on social institutions. Geographical mobility, increasing separation of workplace and residence, and increased participation of women in the labor force have placed even greater strains on the family as the basic social unit of our society. Religious belief, the classic stabilizing source of ethics and values, has been seriously threatened by the rise of science which offers no acceptable substitute. The proportion of individuals expressing the view that religious beliefs are personally very important declined steadily between 1950 and 1970 although, interestingly, there has been a resurgence in such belief since 1970.

Our society has become more complex, more interdependent and, therefore, more vulnerable while its classical stabilizing features have lost their strength.

Complexities of Social Evolution

Most of us have believed that it is the growth of the total economy that best enables improvement in the status of those at the bottom of the economic ladder. The country seems undivided on this issue even though technological advance has not benefited all members of society equally. It is surely not surprising that the flood of material goods produced during the last several decades mostly benefits those with the resources to pay for them. But it is more subtle than that. Consider that "Sesame Street," the educational television program designed to help disadvantaged preschool children catch up to their more affluent counterparts, has actually widened the gap between rich and poor because middle- and upper-class parents encourage their children to watch the program and then spend more time discussing it with them while poorer children tend to watch the trash and violence. It is, I concede, ironic to cite television watching as a description of poverty! However, we do not evaluate our circumstances in an absolute way; we compare ourselves with others. It is the universal access to the communications media, making evident how others live, that feeds unrest here and abroad.

A survey reported by *Fortune* magazine revealed that the cleavage lines in American society have altered dramatically. In the thirties there were sharp divisions along economic class lines, workers versus management and the middle class, with the workers deeply distrusting government and big business and desiring to "soak the rich." Today opinions concerning business, government, tax policy, and inflation are much the same at all income levels. Remarkably, it is the better educated, higher income group that considers the size of big business, per se, to be "a threat to the American way of life." It is the least educated, lower income group who resist social change on such issues as religion, abortion, the role of women, divorce, and the work ethic—thereby confusing traditional descriptions of liberalism and conservatism. The environmental and consumer movements became so powerful in a short period precisely because they arose among and attracted numbers of the educated, affluent young people of our society. Though the latter may deny it, these concerns are not much shared by lower income groups. It may be precisely this lack of connection between the divisions created by individual views of these social issues or mores and the more conventional alignments on economic issues that makes "single-issue" politics so disabling for the political process, threatening the two-party system that has served us so well. Moreover, it must certainly be true that when social unrest arises among the most highly favored members of a society, that society must be in deeply serious trouble.

Looking outside our national boundaries, the advance of technology has begun the process of welding humanity, worldwide, into a single community. Modern transportation and communications have given hundreds of millions of persons, worldwide, increasing acquaintance with peoples elsewhere and the quality of their lives. That this has not homogenized us is evident in the events in Iran, in Northern Ireland, in Quebec, in Vietnam, in "Yankee Go Home" signs in so many corners of the world. Economic interdependence, for resources and for markets, may yet serve as a unifying force, but it also places the greatest possible strain on international relations, and so may dominate international relations for the indefinite future.

A vital element in the outcome of this interdependence is the growth of world populations. The carrying capacity of the planet may, in principle, suffice for a much greater population than that of the moment. Some technologies—those that save lives—drive population growth. Others, largely those that make for income growth, tend to reduce the rate of population increase. But many decades, perhaps centuries, will be required to ease the huge disparity between the per capita incomes of the have and have-not nations. As the populations of the latter grow, they become increasingly vulnerable to the vicissitudes of climate and food supply, increasingly tempted to use their mineral resources as economic weapons, thereby fertilizing the soil in which lie the seeds of conflict. And there remains the specter of superpowers glaring at each other across a developing country with resources that both covet. How large a spark is required for international tension to become international conflict?

Opportunities to Enhance Institutional Performance

Most of us would regard as constructive the state's acceptance of an avowed responsibility to safeguard the economic well-being of its citizens, not to rely altogether on the operation of Adam Smith's "invisible hand." A host of social institutions that did not exist when most of us were young is now in place, designed to protect individuals as well as organizations from catastrophe: social security, unemployment compensation, benefits for the disabled, bank deposit insurance, and federal disaster assistance. It is taken for granted that there should be regulations in support of the health, safety, and well-being of individuals. Although the economic system is currently suffering from burdensome regulation, we must remind ourselves that, for the bulk of our population, life has improved not only because the standard of living is higher but because risks generally have been radically reduced relative to the past, largely because of the intervention of the state.

Political power has been redistributed in the direction of greater equity. Women do vote and hold public office; blacks gained effective voting rights in the early 1960s and have become an important voting bloc. One consequence is the new assertiveness of what formerly were disadvantaged groups, a phenomenon that sometimes makes some of us uncomfortable and has certainly added to the complexity of our society. That should not be interpreted as evidence of breakdown of society, but rather of the opposite. From this perspective, the increased litigiousness of our society, the overwhelming of the courts in consequence of increased social conflict may be construed as an indicator of social progress, however turbulent the way.

There has occurred a dramatic increase in public awareness of how our society operates and of the causes and consequences of the interplay of conflicting groups and forces within that society. It is now taken for granted that social arrangements can be altered for the collective good and that it is appropriate to strive for such alterations as a matter of public policy. It has become a political necessity to manipulate the economy to achieve some balance between the evils of inflation and those of unemployment, with recognition of the differential impact of various policy alternatives on specific segments of the population although, patently, we are not very competent at so doing. Although we have certainly not found the solution to many social ills, in at least some areas the illness has been identified in ways that make solutions imaginable if not yet practically achievable.

The social sciences are new and have not yet forged powerful analytical or predictive tools. Nevertheless, they have begun to contribute to social progress. For example, we now collect statistics on the state of the population and the state of society. Economic indicators are essentially a post-World War II phenomenon; social indicators are still more recent. Only in the past 15 years have there been systematic data concerning indicators of institutional progress such as those I used earlier. Because we now monitor the unemployment rate on a continuous basis, one can study the effects of alternative policies and perhaps learn whether we have been making things better or worse.

A similar case can be made for a host of other measures of the state of the society, the economy, and the population that we currently collect. We could not judge improvements in the equality of opportunity until we begin to collect data on the educational, occupational, and income attainment of men and women, of whites, blacks, and other minorities. For that matter, we could not ascertain the efficacy of public health measures until we had reliable measures of the health of the population. Indeed, it can be argued that part of our feeling about the pervasiveness of social ills stems not so much from the fact that things are worse than they used to be but rather from the fact that we have better measures of just how bad they are.

These data have taught us that social engineering is extraordinarily complex. They have demonstrated that various simplistic social policies don't work very well and we are beginning to identify some of the reasons why. There is increasing evidence, for example, that bureaucracies are extremely resistant to change, willing to subvert policy changes to protect their own positions, that those who are already well placed in society are usually better able to capitalize on new opportunities than those who are disadvantaged because they have more skill and practice at manipulating the system. This increasing sophistication about the way social systems operate makes us aware that improving our social institutions is not going to be easy, but it also points the way to the development of better programs.

SCIENCE AND TECHNOLOGY: TOWARD THEIR INSTITUTIONAL WELL-BEING

The case for science and technology as mitigators of human circumstances is stronger individually than on the institutional dimension, where performance has lagged and risks, most notably that of nuclear war, have accumulated. Let me now turn to the other side of this coin, the social context in which science and technology develop and function. Concern for the social impact of technology has now grown to a point where Jean-Jacques Salomon refers to a loss of innocence for society as a whole (see Chapter 8). Modern origins are to be found in the revulsion of feeling experienced by the scientific community in the aftermath of Hiroshima and Nagasaki. Concern that these weapons not be used again, and objection to the fallout from weapons testing in the Southwest, led in turn to forums that soon turned to discussion of possible hazards of nuclear power. The promoters of nuclear energy shared a vision of the contribution that nuclear power could make to the national future but lacked the foresight to make sure that the citizenry was fully informed concerning all aspects of life in a nuclear age. The seeds of distrust sprouted and are today powerfully evident in reaction to the Three Mile Island reactor episode as the public indicates its suspicion that the utility company, the builder, and the Nuclear Regulatory Commission have all been less than forthcoming, possibly with serious consequences for the future of nuclear energy.

Science and Technology on the Agenda of Public Choice

Like the two-by-four required to get the mule's attention, social protest over the last decade has been unwelcome to those exposed to it. I believe that, in time, we will be grateful to the protesters who forced the nation to give open and complete examination to this potpourri of issues including those surrounding the future of our energy supply. I can be somewhat sanguine that, for that reason, when, one day, we do have a national energy policy, it will rest on a comprehensive analysis of the economic, environmental, and social costs, as well as the benefits to be achieved by the mix of energy sources to which we must ultimately agree. While market mechanisms will be extremely important in determining that mix, I am skeptical that free-market mechanisms alone will optimize the long-term energy policy which will be required or enable an orderly transition to that time.

Debate frequently hinges upon issues of safety: leakage of radiation, disposal of radioactive waste, core meltdown, terrorism, or proliferation of nuclear weapons. But as has been noted elsewhere, arguments about safety frequently are surrogates for arguments about the nature of society itself. Three Mile Island will be seen, however, as legitimizing the safety arguments. Many of those most vocal and articulate in opposition to nuclear power are in fact arguing for the decentralization of American society. They regret the rise of bureaucracy and the growth of corporations. They have adopted nuclear power as the symbol of their disaffection with a materialistic society that in their view will ultimately prove disastrous to both the physical and social environment of man, of their disaffection with the organization and functioning of an urbanized society run by a democratically unaccountable bureaucracy. These are legitimate social issues. What is important is to see to it that in the continuing debate concerning nuclear energy, these issues are not permitted to distort objective evaluations of safety, economics, or technical performance.

Nor will the debate remain limited to nuclear power. Our technological civilization offers a number of attractive symbols which will come under equivalent attack in due course. And it would be well to recognize that, beneath the surface, the environmental and the consumer movements constitute a cry of protest for the sense of powerlessness of the individual educated citizen; again the surrogate for the actual complaint is a discussion of safety. The response has been attempts at examination of such matters by formal risk and cost-benefit analysis. But the environmental problems of our day involve risks and benefits that usually accrue to different groups and costs, risks and benefits that are incommensurable. Costs are reckoned in dollars, benefits in aesthetic or material values, and risks in human lives. It is for this reason that while risk-benefit analysis can certainly inform the decision maker, his decision must necessarily still turn on a value judgment. The acceptability of a given level of risk remains a political, not a scientific question. When scientists enter these lists but fail to recognize the boundaries, unspoken ideological or political beliefs easily becloud seemingly scientific debate.

Maintaining Science as a Resource for Decision

Difficulty arises in the scientific community from confusion of the role of scientist qua scientist with that of scientist as citizen, confusion of the ethical code of the scientist with the obligation of the citizen, blurring therefore the distinction between intrinsically scientific and intrinsically political questions. And yet that need not happen. In the mind of the scientist there need be no conflict between science and human progress. The scientific ethos itself should compel the behavior of the scientist when he contributes to evaluation of the social value of a specific technology. Unfortunately, when presentation of his analysis and recommendations is also suffused with a social or political ideology, the scientist advocate can all unconsciously become a partisan and leave his ethos behind. And we have seen that occur.

What seems lost on some who would participate in the debate on the place of technology in our society, particularly those concerned with possible environmental carcinogenesis by radiation or chemicals—the great concern of our day—is that the necessity for scientific rigor is even greater when scientific evidence is being offered as the basis for the formulation of public policy than when it is simply expected to find its way in the marketplace of accepted scientific understanding. Science itself can benefit by early publication of properly documented preliminary findings. But surely public policy should not rest on observations so preliminary that they could not find acceptance for publication in an edited scientific journal. And yet that has happened repeatedly.

Political decision makers have no choice but to rely on the validity of what seems to them to be the findings of rather recondite science, thereby placing a heavy onus on scientists who bring such matters to attention. Announcement of each experiment in turn generates public alarm that can neither be justified nor assuaged. Once a compound or power plant, or new technology, has been publicly called into question, however meager the evidence, decision concerning its use becomes unavoidable. The sensible guide would be to accept substantial hazard only for great benefit, minor hazard for modest benefit, and no hazard if it can be avoided without penalty. But in most cases to date quantitative assessment of risk is entirely lacking; accordingly, the current guide appears to be to place a value of minus infinity on any possibility of carcinogenesis—a position that cannot be indefinitely sustained.

For most environmental pollutants that have recently been called to attention we are concerned with potential but as yet undemonstrated hazard. Statistically speaking relatively few persons are known to have been seriously damaged by man-made chemicals. The absolute number is, of course, meaningful and deplorable. But as a percentage of total mortality it is minuscule. The assertion that more than 80 percent of all cancers is occasioned by environmental factors refers to the natural, not the man-made, environment.

In any case we have become highly conscious of environmental health problems. A host of institutions, public and private, are alert and vigilant. The result has been a stream of regulations, each well intentioned, most indeed commendable. But in the absence of persuasive data concerning the magnitude of risk to humans, the sum of such regulation can engender public cynicism, ensnarl life in the workplace, and slowly paralyze the economic life of the nation. I applaud the evolution of the Clean Air Act from 1970, when it mandated reduction of risk to zero irrespective of cost, to 1977, when it asked that decision be based on comparison of marginal cost with the marginal benefit of pollution abatement. That returns to the scientific community the burden to identify and quantify risks and relate health effects to exposure levels, as it leaves to all of us the responsibility for developing a meaningful risk-benefit calculus which is now so drastically lacking.

A decade ago it may have been desirable to flag public attention to potential hazards and proceed as if each were a clear and present danger; it is time to return to the ethics and norms of science so that the political process may proceed with greater confidence. The public may wonder why we don't already know that which appears vital to decision, but science and technology will retain their somewhat diminished place in public esteem only if we steadfastly admit the magnitude of our uncertainties and then assert the need for further research. And we shall lose that place if we dissemble or if we argue as if all necessary information and understanding were in hand, whether the question be dietary prevention of atherosclerosis, the health effects of air pollutants, or of food additives, or of microwaves, the economics of solar energy, the properties of waste disposal systems, or the social consequences of electronic mail systems or of linkage between large data bases. Scientists best serve public policy by living within the ethics of science, not those of politics.

These considerations reveal a painful dilemma. All of us cherish the democratic ideal that matters might be so arranged that persons affected by public policy could have a voice in framing that policy, even an opportunity to vote on that policy. Implementation of that ideal with respect to public management of technology and applied science has a less than noble history. Unless opposed by major economic factors understood by the polity, scare tactics prevail all too readily. Witness the outcomes of plebiscites concerning fluoridation of community water supplies. It is facile to suggest that a more scientifically literate population could more readily and successfully make such decisions; the issues that must be factored into decisions concerning nuclear power, coal combustion, automotive emissions, and many food additives are complex even for the practicing scientist. We need no wholesale demonstrations that "a little knowledge is a dangerous thing." Accordingly, we require institutions and procedures that are democratically accountable while informed decision making itself must be left to those to whom we have temporarily given that authority on our behalf.

For two centuries, concern has occasionally welled up for the societal consequences of one or another technology. But, until recently, science itself was viewed by society as an autonomous venture of high integrity whose conduct is best left to the scientists. That wall has been breached, probably forever. With respect to social science the principal public concern has been with the protection of privacy, a concern that also affects a multitude of uses of modern information technology. The Privacy Act which knowingly inhibited the use of the survey instruments dear to sociologists also inadvertently crippled epidemiological research as well. I trust that we can yet repair that defect. In medical research, public concern has largely been for the use of human subjects. A variety of institutional procedures, such as those for informed consent, have been installed. The controls are somewhat cumbersome but worth it, and have elicited little complaint, except for the occasional hypocrisy they may engender.

Safeguarding Scientific Rationality

In other instances, however, the problem has first been flagged to attention within the scientific community, then elevated to the status of a public issue.

There has been pressure not to undertake certain studies because of fear that society might not be able to live with the answers gained in investigations of, for example, racial differences in intelligence, the association of criminality with chromosomal structure, ethnic voting behaviors, and sociobiology. The most recent cause célèbre concerns research utilizing recombinant DNA, the technique in which a fragment of the genetic material of one species is inserted into that of a second. This simple technique evoked a crescendo of concern which has now diminished considerably. Again, alarms sounded with respect to safety were in some instances surrogate for concern for the very success of the research. Some really wished to urge that man should not knowingly intervene in the workings of biological evolution, as if we had not already done so. Others feared that this research might be a major step along a trail that could lead to the capability of genetic manipulation of man himself, concern that led to the position that, as soon as a scientist can see some direct pathway from his present work to an evil outcome at some future date, no matter how remote, that scientist should abandon the field. Sir MacFarlane Burnett, that most distinguished Australian immunologist, said, "It is a hard thing for an experimental scientist to accept, but it is becoming all too evident that there are dangers in knowing what should not be known."

To Dr. Burnett I reply that it must be far more dangerous to live in ignorance than to live with knowledge. He forgets that the uses of science are indeed unpredictable. He ignores the intrinsic value of knowledge of our own genetic mechanisms, the immediately obvious practical applications of research in this field, and the applications which must lie beyond the horizon of our imaginations. Conversely, the ugly possibility that concerns him could occur only at the end of a long and extraordinarily difficult experimental road, in full view of many observers. There will be ample opportunity to prevent the feared ultimate outcome whereas termination of the entire enterprise at this stage denies to posterity all of its potential fruit.

Nor can I easily condone any abridgement of the freedom of scientific inquiry. Historically, freedom of inquiry like freedom of speech, of religion, of the press, and of assembly, came to be cherished precisely as the scientific search for truth freed mankind from dogmatic religious and political thought. Scientific inquiry has challenged the dogma of an authoritarian world for the last 400 years. It has freed men's minds as it eased their toil. It was Thomas Jefferson who said, "There is no truth on earth that I fear to be known."

With additional experience the problem of research with recombinant DNA will surely evaporate. What concerns me is that an unjustified minor abridgement of scientific freedom today could pave the way for major infringements tomorrow and in turn could constitute the first step along a trail which must inevitably lead to loss of those other freedoms that we cherish. Neither law nor tradition confers an absolute right of freedom from all restraints. But we need accept no constraints other than those essential to protect against injury to other values that we cherish. To use the power of government for the suppression of *ideas* that might otherwise flow from research would take us back to an era of dogmatism from which mankind has only recently escaped. And it would be a feckless course. In the long run it is impossible to stand in the way of the exploration of truth; someone will learn, somewhere, sometime.

And again whereas my colleagues and I have been made uncomfortable by these challenges to science, in the end we will be grateful. Science is a public venture. Jacques Monod said, "The only goal, the sovereign good, is not the happiness of man, not even his temporal power or comfort, nor the Socratic 'know thy self'—it is objective knowledge itself. This is a rigid and constraining ethic which, if it respects man as a supporter of knowledge, nevertheless defines a value superior to man himself." Whereas a scientist might be tempted to associate himself with that view, it will not suffice. It may account for the behavior of individual scientists, but it does not impel a society to provide support for the scientific endeavor. That support rather is provided very largely because of the instrumentality of science as the progenitor of useful technologies. Accordingly, the larger society will, increasingly, insist that not only the uses of science but the conduct of science be democratically accountable, however uncomfortable that may render scientists. For better or worse, the terms of a new social contract between the scientific community and the larger society are now being forged.

It remains evident that science cannot be planned very far into the future; that we shall be surprised is the only certainty. The governments of countries with planned central economies plan research and development to accelerate attainment of the societies they already have in mind. We prefer to evolve—to accept the pluralistic approach to our future and to have the multiple instrumentalities of our society, universities, government, and industry, engage in the research and development that they find most appropriate—and then let us all, however awkwardly, choose which of their products to adopt.

The Exposure of Science to Historical Forces

In a speech on "technology and catastrophe" already mentioned by Salomon (see Chapter 8), Harvey Brooks pointed out that the optimists among scientists continue to see scientific understanding as a worthy goal in itself and as the means for expanding the planet's energy resource base by converting otherwise unusable resources, for minimizing human pain and disease, and, in his phrase, "as a means of so managing affairs that the goal of an equitable harmonious world need not be denied to mankind." But he noted that the path is perilous. Success demands an endless stream of greater and lesser appropriate decisions with little forgiveness for error. As he said, "the scope of human choice and freedom widens, at the same time that the possible price of error escalates."

We have little understanding of the factors that prompted the extraordinary acceleration of science in Europe two centuries ago, or of the subtle circumstances that caused the technological torch to move from Europe to America a century later—presumably some blend of the frontier, democracy, primitive capitalism, and the entrepreneurial talents of a handful of inventors. But we do know that we now live in an overpopulated, competitive, interdependent world that the United States no longer dominates and that other once powerful, prosperous, dominant civilizations have disappeared. Even now we are sharing that torch with several other nations.

Please understand that I am concerned for a serious possibility. Few of us gathered here can readily believe that the United States could lose its preeminent place in the world as other nations have within our lifetime. Somehow we feel that America is exempt. Yet to think so is to fly in the face of history. Many nations at the apex of power were inwardly doomed when their willpower began to falter. Therefore, we should be most careful about retreating from the specific challenges of our age, reluctant to turn away from the frontiers of this epoch.

Each noteworthy civilization has grappled with the great problem of its time. For the Greeks, it was the organization of society; for the Romans, the organization of empire; and for the Medievalists, it was establishing their relationship to God; and for the Europeans of the fifteenth and sixteenth centuries it was to master the oceans. For the last two centuries, it has been the scientific understanding of nature and the creation of an industrial society. For tomorrow the challenge is to continue the latter tasks, to determine how mankind can live in harmony on this finite globe, establish permanent relationships to its finite shrinking resources as well as with infinite space, and to enable achievement of the individual potentials of human beings as we reduce the ravages of disease.

If, instead, intent on a risk-free society, we succumb to a national failure of nerve, if we always heed the nay-sayers, then again Shakespeare will have been proved prescient, in his lines from *Julius Caesar:*

> There is a tide in the affairs of men
> Which, taken at the flood, leads on to fortune;
> Omitted, all the voyage of their life
> Is bound in shallows and in miseries.
> On such a full sea are we now afloat, and we must
> Take the current when it serves
> Or lose our ventures . . .

Whatever other policies we follow, I know that we can be unabashedly optimistic concerning the prospects for continuing great discoveries in science. Indeed, in retrospect, a century or two hence, this time may well appear to have been the heroic age of science in all disciplines. Geophysics, astrophysics, meteorology, paleontology, high energy and solid state physics, chemistry and material science, and, of course, molecular and neurobiology are all bursting with dramatic new insights and understandings. Yet we should note that even in science, the international position of this country is changing. For two decades after World War II, the United States built the largest, most powerful, most productive scientific endeavor in all of history. And it has never been more productive than today. But the economic growth of other nations has enabled them also to develop their scientific potential. Today, we constitute perhaps one-third of the world effort, albeit our enterprise still seems somewhat disproportionately successful. But it is clear that the others are rapidly gaining. Since the fruits of science are universally available, we should be pleased, not dismayed by that fact, as long as we maintain our own capabilities.

Our current malaise then stems from a few bad experiences and from the time delay in meeting the high hopes and expectations raised in the minds of those who appreciate the great power of science and the force of technology. Those expectations have taken on a new light as science has also revealed the true condition of man on earth. I see no alternative but to address vigorously the principal questions of science itself and to use our ever-widening understanding and increasingly sophisticated technology with grace, charity, and wisdom. We are not omnipotent but neither are we unwitting foils of powerful forces over which we have no control. Our joy must be found in those acts by which we exercise our unique human capabilities to eradicate what we abhor and to promote that which we value and cherish. For myself I retain my faith that science, which has revealed the most awesome and profound beauties we have yet beheld, is also the principal tool that our civilization has developed to mitigate the condition of man.

III

Human Needs and the Future of Invention

11
Energy
Wolf Häfele

Our workshop on energy was meant to exemplify the theme of the symposium by discussing the energy problem in view of the human prospect. All participants in the well-attended workshop felt very strongly about this relation. We began with a review of energy demand, urbanization, and decentralization, and reflections on forecasts, scenarios, and conceptualizations. We then considered the world economic situation, mostly in view of the problem of developing countries, and then turned to the problem of the prototypical nature of nuclear risks and society's perceptions of them.

IS THERE AN ENERGY PROBLEM?

The energy problem appears as the tip of an iceberg whose main body is the much more general problem of science, technology, and the human prospect, the topic of this international symposium. So it may be appropriate to consider the energy problem as a pathfinder for the more general problem. Do we have an energy problem? The answer is not trivial. It can indeed be argued that there is no problem provided we limit our considerations of energy to those that are purely objective. Objectivity implies depersonalization, which is the very essence of the natural sciences. Any scientific statement or theory must be provable to anyone who knows the rules of conducting experiments, making logical deductions, and, in most cases, applying mathematics. Objective statements about energy must therefore be acceptable to everyone. This degree of objectivity is entirely satisfactory to the physicist who is concerned with the problem of understanding charmed quarks, where science can in principle give him an answer, but it is not much help to the parent who is concerned with the problem of educating his children. It all depends on what, precisely, we mean by the word "problem".

It must be recognized that the energy problem is primarily a problem of perception extending well beyond the domain of objectivity. Every individual or society has a certain perception of reality and that means, among other things, a perception of the energy problem. The uses of energy are deeply embedded in our personal lives and in the lives of societies, and therefore the energy problem most often refers to that very fact and to the way in which it is perceived. Such perceptions can be quite different from person to person or from society to society. In Europe, for instance, the energy problem is perceived in terms of the political security of supply. In the poorer developing countries, it is perceived in terms of the high price of oil. And for some planners and analysts, the energy problem appears as a misfit between the present infrastructure of energy supply and demand on the one hand and an infrastructure that would permit large-scale uses of unconventional fossil, nuclear, and solar resources on the other hand. This list is not complete, as for instance a case can be made for a still different perception by the oil-producing countries. But having all that in mind, we will now concentrate on a few more specific aspects of the energy problem that perhaps will help us to shed light on the topic of this symposium.

Energy Demand

Energy demand is not deterministically related to the economy or to personal consumption. Energy provides services through its various uses. For instance, possession of a running car, an illuminated book, or a warm room are such services. Their nature is qualitatively different from that of energy for they can be consumed. The state of having a car running is consumed by friction losses, the state of having an illuminated book by the absorption of photons, and the state of having a room warm by the dissipation of heat. In all cases, the entropy increases. The fairly deep-seated point is that energy obeys a law of conservation while entropy does not, and this makes for a fundamental difference between energy and the services it can provide.

It is important to recognize that many of the services that are provided by the use of energy can be obtained in other ways. The use of energy can be substituted by the use of labor, capital, and knowhow. A car, for instance, can be kept running forever by having it suspended by super-conducting magnets in a vacuum tube and always operating on a geodetic trajectory. This is not a very practical proposal because it requires excessive amounts of capital, labor, and knowhow, but it illustrates that such mutual substitutability does indeed exist. What we experience therefore is a coincidence of today's degrees of substitutability of energy, capital, labor, and knowhow for providing the services that people and society desire. As a matter of fact, today and worldwide the uses of one watt (in the general caloric sense) seem to correspond to the uses of one dollar of capital stock. The degree of substitutability of labor and knowhow is more difficult to quantify but leads right away into the problem of society. Another word for substitutability through knowhow is productivity, and the interplay between labor and productivity together with capital leads directly to many of today's political problems. A hundred years ago the issue was the interplay between capital and labor which led Karl Marx to write his book *Das Kapital*. Today, we face a situation where a book on energy and productivity may be forthcoming. In both instances, it is the relation to labor which is of fundamental relevance.

Realizing the substitutability of energy, it is therefore quite natural that the forecast of energy demand is highly controversial and often a matter of fashion. In the United States, the present annual energy consumption is close to 80 quads (1 quad = 10^{15} BTU). A few years ago, one could see forecasts as high as 200 quads for the years 2020 to 2030, based on extrapolations of past trends. Today, one sees forecasts that remain at 80 quads by requiring people to adopt a lifestyle where only very small amounts of energy would be consumed. Along with such more extended forecasts to the year 2020 or 2030 goes the necessity for considering energy demand in a globally comprehensive way. Already the problems of supplying crude oil force us into such considerations, but in the future the growth of developing countries will provide still further motivation. It is primarily the growth of the world population which must be faced. Today we have four billion people on earth. It is hard to argue that there will be less than eight billion by the year 2030. In any event, it is a matter of prudence and responsibility to anticipate such a situation. If the miracle happens in which a population of eight billion does not materialize, so much the better.

Today, the average per capita energy consumption for the world as a whole is two kw years/year. A somewhat detailed analysis for seven world regions, which together cover the whole world comprehensively, has led the Energy Program of the International Institute for Applied Systems Analysis (IIASA) in Austria to consider two scenarios for total world energy demand. In the lower demand scenario it turns out that the average per capita consumption increases to three kw years/year by 2030, while in the higher demand scenario the number is five kw years/year. One may consider these to be moderate estimates.

Some other estimates are more extreme, in one direction or the other. There are two schools of thought. One school wants to use the energy problem as a vehicle for their desired change of life-style. Their proposition is to consider 16 tera watts (tw = 10^{12} w) for the year 2030 and not more, thus remaining at an average per capita consumption of two kw years/year as world population grows. Any increase of per capita energy consumption in the developing countries must therefore be accompanied by a decrease of per capita energy consumption in the developed countries including the Soviet Union. This global energy strategy implies severe changes in life-styles in the industrialized countries and of expectations in the developing countries. The other school of thought is found in the countries that are members of the United Nations "Group of 77." They expect continued growth in the neighborhood of five percent per year for the world as a whole, with higher values for the developing countries and somewhat lower values for the industrialized countries. Along with such growth rates, studies for the United Nations Group of 77 expect high energy demands.

At IIASA, we concluded that a more sustainable rate of growth would be closer to three percent, so we did not follow the high growth rate of the United Nations Group of 77. On the other hand, we did not follow the very low estimates either. At least for some decades to come, the growth in developing countries is coupled to the growth in industrialized countries. A crude rule of thumb says that in developing countries growth cannot be more than 1.5 percent higher than in the industrialized countries. At IIASA, we therefore hesitated to consider scenarios which imply a positive growth rate of the energy per capita consumption in the developing countries and a negative growth rate for the developed countries, as this implies a complete decoupling.

This is not the place to go more deeply into it. But it is obvious that expectations for future life-styles, both in the North and the South of the world, together with the forthcoming growth in world population, lead deeply into the societal domain and strongly determine perceptions of what the energy problem is all about. These cannot be the objects of an objective scientific assessment alone. The energy problem is not a problem that can or should be completely depersonalized. What then is the impact of the energy problem on the conduct and thereby the nature of science in our days? It appears that the point of this symposium is well taken.

Urbanization and Decentralization

The notions of urbanization and decentralization point to another aspect of the energy problem. It is largely through urbanization that development has taken place. Research at IIASA has led to the thesis that in conurbations and urban areas the energy consumption density has a tendency to be constant at five watts per square meter, irrespective of whether the status of the society is developed or developing. This result is observed when one considers the natural boundaries of urban areas, which can be different from the political boundaries. In this concept, development means increasing the available urban space with its associated urban population and energy density. It is recognized at IIASA that more work is required, but let us here and now assume that this thesis is true. Let us further assume that development continues to progress through urbanization. At 500 cap/km^2, eight billion people and five w/m^2, this then leads to a total energy demand of 80 Tw years/year for complete urbanization, or ten kw years/year per capita. While this is merely an order of magnitude consideration, it nevertheless confirms the range of total energy demand considered above. In the light of this completely independent, purely spatial consideration, our previous figures of three to five kw years/year make sense.

But there is a school of thought which maintains that it is exactly this trend of urbanization that must be broken; that in the developed countries a decentralization should take place and in the developing countries a radically new mode of development should be invented which does not go through urbanization but takes place in the rural areas. While this is primarily a political and sociological goal, it nevertheless articulates itself in the language of the energy problem. Accordingly, it is postulated that such decentralization should go in parallel with the uses of renewable energy sources and a strong decoupling of energy and the economy. This is called the "soft path" as opposed to a continuation of the trends and relations of the past which is called the "hard path."

Renewable energy sources in this context are wind, biomass, hydropower, and the local uses of solar power. These sources have a supply density between 0.1 and one w/m², a fair number being 0.5 w/m². Now let us assume the two kw years/year of the new life-style, the eight billion people of the year 2030 and a supply density of 0.5 w/m². This leads to an area of 32 million km², the area of the United States and the Soviet Union taken together, from all of which the renewables must be harnessed. At three or five kw years/year, the required areas would be 48 or 80 million km². One may keep in mind that the total habitable area of the world is not far from 80 million km². Again, this is not the place to go into greater detail. It is obvious that the problem is by far more complex, but with all the necessary caveats this brief consideration again throws light on the deep political and societal implications that can result from different perceptions of the energy problem by various individuals and groups. Again, we have arrived at the topic of this symposium.

Forecasts, Scenarios, and Conceptualizations

Today as in the past there is a great deal of forecasting. There are important fields where this has to be done and where the appropriate methodology has come to hand. Major industrial investment decisions require forecasts of costs, sales, and prices. Every budget of a state requires a forecast of expected revenues. There is no doubt that the object of such forecasting is a perception of actual future reality in the full sense of the word. The methods in use are of a causal or statistical nature but they all amount to trend extrapolations of a more or less sophisticated type. The longer the time horizon, the more difficult it is to forecast successfully. Most methods are capable of looking ahead only a few years. But some aspects of the energy problem have a much longer time horizon. At IIASA and elsewhere, it is considered appropriate to envisage a time horizon of 15 to 50 years from now, extending to the year 2030. Any major change in energy infrastructure, such as a transition from today's infrastructure of energy demand and supply to a new one that permits the engagement of the essentially unlimited resources of nuclear and solar power, requires at least 50 years for its accomplishment. This is not the place to elaborate in greater detail, which has been done elsewhere. The point here is that existing methods are not appropriate for forecasts over such an extended period. Yet responsible action requires an approach which is as much as possible rational and in that sense scientific.

A partial solution to this dilemma is to reflect on as many as possible necessary conditions that must be fulfilled for any kind of future. For instance consumption cannot be larger than production, investments must be accumulated before production can start, and resources can be consumed only once. While it is obvious that imposition of these conditions still leaves many choices, it nevertheless reduces the number of choices that otherwise might appear feasible. One calls this type of approach "scenario writing." The more elaborate and comprehensive the scenario, the larger the number of necessary conditions that must be met. Often as a result the problem in question can be roughly quantified, in that upper and lower limits can be identified. The above-mentioned three and five kw years/year limits for the per capita global energy demand of the year 2030 are the result of such an exercise. Because the scenario writer must explicitly identify what he considers to be a necessary condition, this in itself leads often to clarifications.

When talking about the energy problem within the horizon of the next 50 years, it is necessary to consider the global nature of the problem. And that simply means that resources can be consumed only once, a necessary condition as observed above. And yet it is true more often than not that energy planning for a nation is done with great care and attention to internal demand, then pointing to certain import requirements and simply leaving it at that. The often unconscious implication is that the world is somehow infinite. But we must consider the energy problem in a finite world, where different nations cannot import the same resources. This condition of global comprehensiveness reduces to a large extent the number of conceivable energy futures.

No scenario approach can be considered a prediction of the future. In addition to necessary conditions, there are sufficiency conditions and we must be very explicit that we do not know the sufficiency conditions for predicting the future. We will probably never be in a situation to predict the long-term future. The distinction between a scenario and a short-term forecast, which indeed is often meant as a prediction for the short-term future, is therefore significant. And many analyses are better viewed as scenarios than as forecasts.

This distinction is usually not perceived by society. There has been and often still is a deep-seated conviction that natural sciences can predict things in a deterministic way and, needless to say, the natural scientists have contributed significantly to that impression. This conviction is enhanced by the use of numbers. In the context considered here, numbers can have two distinct meanings. When they articulate the result of a physics, chemistry, or engineering analysis, they mean exactly what they say, for example that the velocity of a particular bicycle is 3.4 m/s, not more and not less. The basis of such rigor is the existence of laws of nature that are synonymous with a mathematical relation. When numbers are meant as a quantitative expression of a qualitative state, they can have another meaning. The above-given number for the energy consumption density in cities being at five w/m², for example, does not mean that this number is precise or that it follows from a law of nature whose mathematical formulation is known. Instead, it means that the best estimate of the density is not 2.5 w/m², nor is it ten w/m², but in between, closer to five w/m². The appropriate perception of quantitative analyses that are more scenarios than forecasts, and the appropriate assessment of their meaning in operational terms, is difficult for both scientists and society. Talking about the difficulty is comparatively easy, but doing it right in an actual situation is exceedingly complex.

There is more to it. The necessity for distinguishing between forecasts and scenarios, and for distinguishing between the meanings of using numbers and other related considerations, all point to a limited scope for scientific analyses and ultimately for science itself. We more and more realize that it is not only the unknown but also the unknowable that makes the future unpredictable. Is that a temporary shortcoming or is it a principal feature of reality? Probably most of us will agree that it is a principal feature of reality. But why then make any analysis of the energy problem or any other major problem of society if the future is in principle unpredictable? Bearing that question in mind, it is useful to reflect on how we conduct our personal lives. We always make plans, we always have hopes, we always expect something to happen and anticipate a certain sequence of events. When we are young we think of having a certain training and then a professional career; we expect to have a family and to raise children. I would like to call those conceptualizations. We conceptualize in spite of the fact that we know that the future may turn out to be radically different. As a matter of fact, it is probably the very essence of human life to conceptualize. But it must be observed that the conduct of a human life usually requires prudence and wisdom too. This is necessary if we are to accept and thereby master the unexpected. When conceptualization, prudence, wisdom, and many other human activities are all in balance, there results a rich human life.

I now take the liberty to generalize the attributes of our personal lives into our societal and political life. The first observation is that, as in our personal lives, conceptualizations are also necessary for the life of society. And equally these social conceptualizations should not be confused with objective scientific findings. It is disastrous when objective scientific findings are understood as conceptualizations, and it is even more disastrous when conceptualizations are presented as objective scientific findings. Instead, one should distinguish among hard science, forecasts, scenarios, and conceptualizations. There can be much mutual intertwining, but the distinction should be clear.

Implicit in this observation is a recognition of limits to the realm of science. Then we can recognize that it is not so that we will know with certainty if we just wait a little longer until science has done the remainder of its job. Instead, the realm of science is a finite one when it comes to making decisions for real life, when our future is at stake. While it is mandatory to use the maximum scientific rigor and to employ our intellect to the fullest possible extent, it is equally mandatory to reinstitute the realm of prudence and wisdom and other human dimensions of individual and of public life. Unfortunately, our modern societies have no institutions for this purpose. Where does one go to find wisdom? Only when those human virtues are taken together and when our institutions are again able to reflect them can one hope to solve the big problems that are ahead, the energy problem among them. It is not an objective scientific problem alone. It is a problem of life and death. Again, we have arrived at the topic of this symposium. Its importance cannot be overestimated.

ECONOMIC GROWTH, RESOURCES, AND ENERGY REQUIREMENTS

Bruno Fritsch, Professor of Economics at the Federal Institute of Technology in Zürich, reviewed prospective energy supplies in the economic context. Physically the energy resources available are nearly unlimited. The earth-atmosphere system receives a total input of 175,000 Tw from the sun, of which man's requirements in 1977 were less than eight Tw, the 22,000th part of the solar energy input. Apart from the sun's energy, the following sources are, in principle, available to men:

fission (FBRs)	5×10^6 Q
fusion	10×10^6 Q
coal	200 Q
oil	16 Q

$$(Q = 10^{18} \text{ BTU} = 2.52 \times 10^{17} \text{ kcal})$$

If energy appears to be scarce today it is not because there are limits on its sources but because we have developed a disequilibrium between the investments which convert potential into actual energy supplies and those which determine energy-consuming economic growth. This kind of disequilibrium has emerged as a result of holding the price of energy below the efficiency level for some 30 years.

Economic growth not only requires more inputs of energy, but must also make energy available by providing resources for the installation of generating systems. For any given technology, there exists a fairly precise relationship between the rate of economic growth and the rate at which such generating systems can be set up. The energy sector of the economy *cannot* outgrow the nonenergy sectors, and vice versa. The power, in watts, required per dollar per year varies from 0.3 to 2.5, with the average (1977) at 1.3 watts per dollar per year.

Assuming that the world population in 2020 will be eight billion people, and that the average per capita income will be $3,200 (at today's prices), the world GNP would be 25.6×10^{12}. Assuming an average energy intensity of production of 0.6 watts per dollar per year, which implies both savings and efficiency gains, then the world energy demand in 2020 would amount to 15.3 Tw.

Even if fossil sources were still available, they would be unequal to demands on that level, because limits will be set by emissions of CO_2 and SO_2 (for coal), as well as by the limited availability of water and transport facilities. We cannot supply future energy requirements of 15 to 20 Tw from fossil sources alone. Nuclear as well as (hybrid) solar energy is an indispensable link in the chain of energy sources on which we must rely for the future. It is necessary to allow investment lead times of 15 to 20 years in planning changeovers from one energy source to another. Anticipating the age of substitutability in materials, capital and energy requirements will have to be primary elements in the structure of economic systems.[1]

RADIOACTIVE WASTE: A PROTOTYPE HAZARD

H.W. Levi of the Hahn-Meitner-Institut für Kernforschung, Berlin, dismissed as impractical the demand that billions of people should support themselves from the resources of the earth and still leave the earth as they found it. There can be no large-scale offer of usable energy without leaving lasting footprints on the earth.

Radioactive waste is presently the one long-term consequence man's life on earth will have which is in everybody's mind. It is one of the most heatedly debated issues in the nuclear controversy which, in fact, is a technology controversy. Though nuclear power is by no means the only human activity which will leave lasting hazards on earth, it is probably the one which made people for the first time aware of the existence of hazards of this kind. As a consequence, the hazard associated with radioactive waste has become a prototypical hazard.

Nuclear waste will remain hazardous for long periods. High-level waste will remain a significant hazard for a time period in the order of 1,000 to 10,000 years. No doubt, this is a period beyond any predictability of the technological, political, and social state of the human society. The consequence has to be isolation of radioactive waste from man in a way that no human activity is required to maintain this isolation. Along this line, a general concept of final disposal has been developed internationally, relying on the time scale of geologic rather than on that of human systems.

The essentials of this concept are storage of chemically and mechanically stable waste forms deep underground enclosed in a stable geologic formation keeping circulating ground water away from the waste. The main safety concern about the geologic containment is to essentially maintain the long-existing geologic equilibrium even upon mining the repository and charging it with heat-producing nuclear waste. The geologic formation containing the waste repository is the first and basic barrier between waste and man.

The repository's depth provides a second barrier, which would become effective if the geologic containment failed. Radioactive species dissolved in moving ground water would then have to penetrate huge masses of rock and soil with a high sorption capacity. This would drastically delay their migration and their appearance in the biosphere. The stable waste form is a third barrier whose leach resistance would under certain conditions further delay the transport of radioactivity from the repository to the biosphere. There is hardly any doubt that the waste can be safety isolated in a geologic repository for at least 10,000 years. Further generations will be able to live quite comfortably with this hazard, probably more comfortably than with several other problems we are going to leave them. It is therefore most remarkable that large segments of the public are nevertheless worried about the 10,000-year safety of waste repositories. It is even more remarkable that there are many people worried about safety in terms of millions of years, that it might not even be justified to neglect the very small hazard remaining after 10,000 years which is in the same order as that of naturally occurring radioactive minerals.

This is a new dimension of public fear. It reaches far beyond the life span of the individual concerned and it reaches also beyond the life span of any descendant to whom an individual may have an emotional relationship. It is kind of a depersonalized long-term fear closely related to this impractical demand to leave the earth as it was. If this extreme attitude were to develop toward other long-term technology hazards it could eventually jeopardize the use of any energy source. In the specific case of radioactive waste disposal this public attitude already hampers any reasonable technological development in the field, including adequate exploitation of an energy source that badly needs to be exploited if we want to maintain our present standards of technology and economics. On one hand, the urgency to put radioactive waste to rest in a final repository is usually much exaggerated as the waste may be safely kept in engineered storage for decades. On the other hand, operation of prototype repositories to demonstrate the feasibility of this technique is by no means encouraged. On the contrary, operation of those facilities is usually considered tantamount to routinely disposing of the waste from a fully nuclear economy and is treated accordingly by regulators. Thus, we are rather deadlocked between two extremes. We neither may use nuclear power and be relaxed about waste waiting safely in engineered storage facilities for a final place nor may we adequately push the demonstration of geologic waste repositories.

The priority given to far-future problems as compared to more actual ones needs careful and thorough analysis. Radioactive waste may stand in this analysis as a prototype of this sort of hazard. The time scale and the complexity of consequences reaching far into the future make a technical analysis of their significance more difficult than for more actual technical hazards, such as nuclear power plant accidents or dam disasters. Therefore, the perception of long-term risks will inevitably be more qualitative, more uncertain, and consequently likely to be more emotional than that of a short-term risk.

It is striking that those fighting for a better world are often more concerned with evils located far away than with problems in their own neighborhood. Following this line it sounds logical that those fighting for a better world in terms of technology may be more concerned with hazards threatening generations living in the middle of the next millenium than with much more actual risks. This way of thinking may be quite dangerous for our society. It may lead us to accept serious risks for the next generation, just to make sure that human beings living 10,000 years from now can benefit from conditions we think they will find comfortable.

There are certainly widespread unspecific fears in human beings who face a highly technical world they cannot fully understand. This fear must materialize in specific objects. Unconsciously, it may be more agreeable to project this fear to the welfare of future generations than to our own. In other words, being concerned about long-term hazards may be perceived as a more ethical form of technology concern. But is it practical to base ethics on a feeling that society has a general right to live with the lowest feasible risk no matter what the costs turn out to be?

Radioactive waste, as a prototype of changing social judgment, should direct our attention to a crucial point. Eliminating all risks from human life is impractical because society has neither financial resources nor the manpower to do so. A systematic reduction of risks, however, is an outstandingly important goal. It is obvious that this goal can effectively be pursued only if priorities are defined rationally and if they are accepted by society. Rational priorities are priorities according to cost-benefit relationships. If we are not able to reach consensus on this very point it might happen one day that we have spent much of our resources to shoot radioactive waste safely into the sun but that we have done very little to reduce the number of deaths on the streets.

To make people approach risks more rationally is one of the most urgent goals we have to work for, if we want to continue to enjoy the benefits of technology. As far as energy is concerned, conventional energy sources will otherwise take advantage from their long familiarity and will probably continue to be more or less accepted. Any new energy source, however, will face the kind of acceptance problems that nuclear fission presently does, as soon as it has reached the same degree of maturity and the hazards become clearly recognizable.

DISCUSSION

We sought to keep the human prospect in view throughout our lively discussions. We sought first to determine the extent possible for the factual basis of the energy problem and then, on that basis, to formulate assessments and policies. To that end two scenarios were envisaged that covered the period up to 2030 and identified a balance of energy demand and supply in seven world regions. The point was to envisage the strong regional interactions in the world as world populations grow further and as general economic development continues, particularly in the developing countries. The size and composition of energy supplies envisaged for 2030 in the two scenarios indeed invite concern. While feasible in principle they point to significant societal problems.

The potential of renewable energy sources was debated with particular attention to hydropower and wood. It became apparent that their potential is significant although limited. By contrast that is not the case for solar power when envisaged as a large-scale hard technology. Contrary to a widespread belief, the land requirements are large but not excessive. Instead the first bottleneck will be requirements for energy storage and materials such as iron and concrete. Later on the extent of the land required will be an active constraint.

The participants made a very strong plea for the enhanced development of a cheap photovoltaic cell and based on that the development of power plants in the giga watt (Gw) domain. Naturally, the question of centralized versus decentralized energy supplies was discussed. Nobody denied the importance of the contribution of decentralized energy supplies. But looking at the global scenarios for the year 2030 made it plain that while there is room for choices, all sources must be engaged and the lion's share of the supply up to that time can come only from fossil and nuclear sources. So it was recognized that the expected contribution of nuclear power continues to be large. Any power generation implies risks, But it was the advent of nuclear power that brought public recognition of this fact. There is a difference between risks and risks as perceived, and particularly so for nuclear power. It may be fair to report that both the risks of going nuclear and the risks of not going nuclear were fully recognized. There seems to be no quick remedy for this.

Another point that was discussed was the problem of the size of capital investments and the lead times involved. There was agreement that in the decades ahead the investment share going into energy might double—not an insuperable obstacle but significant; but an even greater problem seems to be unusually long lead times. It was observed that it may be possible to muddle through to the year 2000, but thereafter the scenarios seem to require more rapid and decisive responses. Intermittently and especially toward the end, the question was raised whether all this energy demand is really unavoidable. In its opening phase the workshop had recognized that there is in general no deterministic relation between economic and private activities and energy demand. It is possible, at least partly, to substitute the services of capital, labor, and knowhow for energy. But the time constants for such substitutions are usually on the order of decades. It may be prudent not to rely on such substitutability very much. The most promising route for a remedy seems to be more knowhow, which, to an extent, means more technological progress. But this should not be alone as an alternative to feasible reductions in energy demand and new approaches to conservation. Prudence requires us to have an open mind in all directions. Thus the dialogue between the proponents of various plans of action should continue.

NOTE

[1]*Science* 191 (1976): 683.

12
Technological History and Technical Problems
Thomas P. Hughes

Technology and science policy makers could turn for general guidance to the history of technology and science, not with the expectation that history is about to repeat itself, but because situations similar to those confronting them in the present can advantageously be studied. Politicians and diplomats are accustomed to look to history for understanding and insight, as their frequent historical allusions show. It may also be argued that phenomena as complex and historically rooted as science and technology cannot be understood in strictly contemporary terms. Our exercise has been to describe an historical situation in the evolution of electric power in hopes of conveying to policy makers the similarity it bears to present-day problems. Let the decision maker judge the persuasiveness of the analogies we set out to draw. History and policy have this in common. Both depend upon difficult moves up the ladder of abstraction from the specifics of trends, confluences, and events. History is content to serve the art that goes into policy making.

Policy and History in Technology: Has the Past a Future?

Since the turn of the century in technology and since World War II in science, large-scale organization has become increasingly common. Science and technology have become highly institutionalized. The momentum of coordinated ideas, expertise, organization, and professional concerns increases the force of history, with which we deal.

This chapter, then, is about technology with momentum, especially electric light and power systems. Because the word technology is used so differently, we should pause to say how it is used here, as a complex system of interrelated factors with a large technical component. Technical refers primarily to tools, machines, structures, and other devices. Other factors embedded in technology, besides the technical, are the economic, political, scientific, sociological, psychological, and ideological. Technology usually has the structure of a goal-seeking, problem-solving, open system. An example of a technological system is an electric light and power system that incorporates economic objectives such as efficiency; political constraints, such as regulatory legislation; sociological organizations, such as business corporations; scientific knowledge, such as that produced in research laboratories; and psychological components, such as the characteristics of influential personalities involved in the system. Electric power systems also exhibit acquired characteristics caused by historical trends and confluences, such as demographic and price trends, wars, and depressions.[1]

Factors, or components, in the electric power systems, as in other technological systems, interact. Furthermore, control of varying degrees of effectiveness and precision can be exercised over the system. Factors influencing the system, but outside of its control, can be labeled—following systems science terminology—the environment. The environmentalist movement, for example, is now an environment for electric light and power systems. (If the goals of the environmentalists are incorporated into those of the electric light and power system, then the environmentalist will become a part of the system, as did some bankers and politicians earlier.)

This chapter is also about problem solving and conflict resolution in the realm of technological affairs. The essential argument is that today we often confuse technical and technological problems and that conflict about technological issues is often misconceived as conflict about technical ones. (I contrast technological and technical following the definitions given above: the technical—it should be recalled—is a component in a technological system.) I shall also argue that technological problems and conflicts are not solved or resolved until they are correctly diagnosed and responded to as technological rather than technical.

As an example of the misconstruing of the technological as simply technical, a recent statement by Philip Abelson, Editor of *Science*, will serve. Referring to the energy problem today he wrote: "The sad part of the situation is that technology and resources exist to enable the United States to live smoothly through the transition to more efficient energy use and long-term and renewable energy sources. But continuation of present paralysis invites turmoil."[2]

A compilation of similar complaints, especially from advocates of particular technical approaches to the energy problem, would be exceedingly long. In essence, many of the writers are saying, "Here is a likely solution. Why the paralysis? Why in the name of common and good sense does society not get along with the development and implementation of it?"

In the past, others, especially engineers and managers, have also been perplexed repeatedly by society's failure to adopt available technical approaches to problems. In the 1920s William Ogburn and other sociologists referred to this failure as societal lag.[3] Marxists refer to an antiquated superstructure resisting the influx of developing technology. Some engineers explain the footdragging of society by a pejorative reference to "politics," and scientific managers describe the situation as irrational. The reason technical responses to the energy problem are not sufficient is that it is technological in character. One recalls a pungent remark by historian Peter Gay who said of radical students in Weimar, Germany that they were offering political solutions to nonpolitical problems. Part of the exasperation surrounding the energy problem stems from offering technical solutions to problems that are broader in nature.

The argument that the energy problem is not simply technical in nature is found in Amory Lovins' well-known article in *Foreign Affairs*. He wrote:

> The barriers to far more efficient use of energy are not technical, nor in any fundamental sense economic. So why do we stand here confronted, as Pogo said, by insurmountable opportunities? The answer—apart from poor information and ideological antipathy and rigidity—is a wide array of institutional barriers, including more than 3,000 conflicting and often obsolete building codes, an innovation-resistant building industry, lack of mechanisms to ease the transition from kinds of work that we no longer need to kinds we do need . . . and fragmentation of government responsibility.[4]

To this the historian might seek to add a sense of "depth in time," as Lynn White has called it, a failing he finds common in persons trained in engineering and social science, leading them to "take a flat contemporary view of phenomena." To White they suffer from an awareness "of the complexity and ramifications of the effects of technological changes in the past."[5] We seek in what follows to describe the emergence of a technical problem and the technical, partial solution found for it. But a variety of nontechnical problems then arose to frustrate the implementation of the simple solution. Then the problem was reapproached as more broadly technological, and it was overcome by organizational, political, and other responses. We hope to evoke a sense that the present—and therefore the future also—does have a past and that it is worth knowing.

A Technical Solution in the 1880s. During the closing years of the nineteenth century, technical journals kept readers informed of the latest developments in a technological affair called by engineers, not usually given to employing dramatic metaphors, "the battle of the systems." Thomas Edison had introduced the world to central-station incandescent lighting in 1882 when his Pearl Street station began supplying customers in New York City's Wall Street district. Edison's station was heralded in America and abroad as an inventive triumph. It was a major engineering achievement as well, for Edison had designed and developed related components for an entire system of electric lighting. Edison-type central stations appeared in leading American and European cities, including Milan, London, and Berlin, all using direct current. Battle lines were, by 1888, clearly drawn to meet the challenge of alternating current. The intensive publicity and marketing of the Edison system foretold that the battle of the systems would not be won simply by technical advantages.

The Edison system had a major technical flaw, soon revealed—the extreme expense of distributing electricity at low voltage. Edison used low voltage for distribution because the filament of his incandescent light demanded 100 volts at the point of consumption. Since transformers could not be employed with direct current, the voltage could not be raised for transmission and reduced for distribution and consumption. If the transmission voltage could have been raised and the current reduced as a consequence, then the power loss in transmission would have been less, for the loss is proportional to the quantity of the current squared. Edison stations, however, were flawed; distribution was limited to a radius of about a mile in densely populated areas.

Lucien Gaulard, a brilliant young Frenchman residing in England and his business partner, John D. Gibbs, showed how the technical flaw in the Edison system might be corrected. Gaulard and Gibbs demonstrated alternating-current transformers that could raise voltages for economic transmission and lower them at the point of consumption. The technical journals publicized Gaulard and Gibbs' installations in England and abroad, and many engineers saw their demonstrations. In 1883, they lit four stations of London's Metropolitan Railway with incandescent and arc lights, using transformers connected on a 16-mile circuit. Alternating current flourished in years following, but Lucien Gaulard did not. In 1888 he died young in Sainte Anne's "madhouse" in Paris, brokenhearted, "a victim of science and his own greatness." His tragic collapse had undoubtedly been precipitated by protracted patent litigation and the loss of his invention to others who developed it far more successfully.

Several years earlier at Turin, Gaulard had discussed his transformer system with Otto Titus Blathy, a Hungarian engineer who worked for the Hungarian manufacturing concern, Ganz and Company. Ganz bought a Gaulard and Gibbs transformer and in the winter of 1884-85 Blathy, Max Deri, and Charles Zipernowski, Ganz's engineers, made major improvements in it. Their Ganz transformer is considered by some as "the first commercial, practical transformer"; another describes it as a major improvement upon that of Gaulard and Gibbs. The Ganz and Company system was widely adopted. By 1890, nearly 70 central stations of various sizes had been placed in operation using the company's alternating-current generators, transformers, and controls. These stations supplied 100,000 incandescents and 1,000 arcs.

In America, George Westinghouse, already famous for his invention and manufacture of the air brake, had also ordered several of the Gaulard and Gibbs transformers. Westinghouse assigned William Stanley and several engineers to develop the transformer system and in January, 1886, formed the Westinghouse

Electric Company, which purchased the Gaulard and Gibbs rights for $50,000. Westinghouse's manufacturing and marketing of the system was impressive; by September 1887, the company had in service or under construction central stations with capacity for 134,000 16-candle power incandescent lamps. By November 1890 the number had risen to 700,000. It should be recalled that Ganz and Company had 100,000 lamps connected by 1900.

The attractiveness of alternating current as a technical response to the distribution problem was further enhanced by the correction of a major flaw in the AC system—the lack of a practical motor. In 1888 a brilliant young American immigrant from Croatia, Nikola Tesla, in a paper read before the American Institute of Electrical Engineers, announced his invention of a motor that would, in his opinion, fulfill the pressing need of the alternating-current system. In July 1888, Westinghouse purchased the Tesla patents for $20,000 cash, $50,000 in notes, and a royalty of $2.50 per horsepower for each motor. Tesla's patents covered a system of generators, transformers, and motors. While the Gaulard and Gibbs, the Ganz and Company, and the early Westinghouse systems were single-phase alternating currents, Tesla's system involved two or more alternating currents out-of-phase.

If technological problems were solved by technical innovations alone, then surely the "battle of the systems" should have ended, but instead it intensified with a series of unorthodox skirmishes. Both Thomas Edison and George Westinghouse became personally involved in the struggle. The details—not always edifying—have been recounted elsewhere. At Edison's laboratories his staff demonstrated the dangers of high-voltage alternating currents to humans by electrocuting horses. A certain Harold Brown, representing the direct-current advocates, challenged George Westinghouse to trial by electricity; each was to hold on to the terminals of the generator he advocated as the voltage was stepped up, and the first to let go would lose the trial and reveal the dangerous character of the current he was advocating and holding onto. A particularly reprehensible episode was the advocacy by the direct-current forces of alternating currents for capital punishment and resulting electrocution of William Kemmler, alias John Hart, in New York State Prison in 1890. The machine used—the public was informed—generated the horrendous alternating current that Westinghouse wanted to run through the streets of America. The Edison Company also made a determined but unsuccessful effort to have the Virginia and Ohio legislatures outlaw currents above 200-400 volts. Edison testified at the hearing in Richmond, Virginia, but his prestige was overshadowed by the Southerners' distaste for northern corporations attempting to influence southern culture.

A practical transformer, polyphase current, a practical polyphase motor, and high-voltage power transmission—there were practical demonstrations in 1891—together formed a system of electrical supply that remedied the major technical flaws of the Edison system and also matched its capacity to supply both power and light. The "battle of the systems," however, was yet not over.

Existing direct-current utilities in densely populated and industrialized areas continued to expand to meet load increases. Their unamortized investment in the direct-current system was too large to permit them to replace it with a polyphase system. To supplement the existing direct-current system with an alternating current system would have meant the loss of the advantages of the scale and diversity of a single system. Furthermore, Edison General Electric, the equipment manufacturing company, was committed by capital, patents, plant, and expertise to direct current. Thomas Edison and other engineers and managers were biased by experience and expertise to the older system. The Westinghouse Company, on the other hand, had vested interests in alternating and polyphase current. The further implementation of the technical solution, then, was frustrated by the way direct current had already been institutionalized and built up a change-resisting momentum. Manufacturers, utilities, and consumers all reacted cautiously until the future line became clear. Contemporaries lamented the irrationality of the situation.

The Technological Resolution. The swing of Edison General Electric to the new way signaled the beginning of the end. The reasons were complex. For one thing, Thomas Edison, a bitter foe of the new current who supported efforts to discredit and outlaw it, released control of the Edison enterprises in the late 1890s. Unfortified by his monumental prestige and conservative influence, the dogmatic opposition to AC in Edison General Electric waned. His withdrawal also cleared the way for a merger with Thomson-Houston, a rising manufacturer of electrical machinery that carried a line of both direct- and alternating-current equipment. After the merger in 1892, the new company, General Electric, introduced equipment for polyphase systems.

A new wave of young engineers also helped make the transition from direct to alternating current. Edison and most of his electricians and mechanics could solve DC problems with elementary mathematics, but Francis Upton was the only Edison pioneer known for his mathematical ability. Alternating-current circuits and equipment brought problems requiring a mathematical sophistication far beyond the competence of Edison and others without formal training. It is possible that Edison's adamant refusal to consider seriously alternating current may have arisen from his inability to cope with it intellectually.

Charles Proteus Steinmetz, by contrast, led the new wave of engineers able to analyze alternating-current circuits. Steinmetz had completed a doctoral thesis in theoretical mathematics at the University of Breslau in 1888 before he left Germany seeking political asylum first in Switzerland and then in the United States. He became a member of General Electric's calculation department in the year of the merger with Thomson-Houston and during the transition to AC machinery. During his first year with the company he presented a paper that introduced complex number algebra as a powerful method for solving AC circuit and machine problems. Steinmetz and the other engineers with AC expertise not only offered problem-solving competence, but were psychologically committed to the transition. The engineering schools soon thereafter contributed to the transition by introducing electrical engineering courses with ample mathematical components.

Another 1896 organizational response came with a patent-exchange agreement between General Electric and Westinghouse. The engineers in each company and especially those in the utilities who bought generators, motors, and transformers had found the denial of obvious improvements aggravating and welcomed facilitation of rational technical exchange. The opposition of direct-current utilities, however, remained a major frustration. The transition for these enterprises came through a combination of events including several technical inventions. During the period 1893-95, as historian Robert Belfield has pointed out, engineers at the Westinghouse Company and at the Niagara power project conceived of a general system of electrical supply integrating direct and alternating current.[6] The inventions used to create the synthesized systems were synchronous generators and phase converters. With these technical couplers, existing direct-current systems and various alternating-current systems could be connected and the advantage of scale and diversity obtained. Utilities that had heavily invested in direct current could phase it out as depreciation dictated. Direct-current stations, no longer economical, were converted into substations where polyphase current, transmitted at high voltages, stepped down by transformers and changed by synchronous converters, emerged as direct current for distribution. In time the direct-current load was changed over to three-phase alternating current. This gentle transition systematically melded organizations as well as technical components. Large American cities, such as Chicago, formerly supplied by several electric utilities—the oldest direct current and newer ones alternating—were now supplied by a large and financially strong organization that absorbed or merged with the others. In the case of Chicago, Samuel Insull's Chicago Edison Company acquired before 1907 20 Chicago utilities with their francises. A culminating merger with the Commonwealth Electric Company established the city-wide utility, Commonwealth Edison.

London's Problem. Unlike major American cities, London was unable to make the technological transition. The failure of London during the first two decades of our century is a striking reminder that technological problems are not simply technical ones. The modes of transition stood as ready for application in the great English metropolis as in Chicago, New York, and Philadelphia. Charles Merz, one of England's most imaginative and bold consulting engineers, had before him the example of the unified system evolving in Chicago when he tried to win over London managers, owners, and politicians to the technological transition that could rationally combine the crazy patchwork of frequencies, voltages, currents, and competing utilities protecting their parochial franchises. Citing his friend Insull, Charles Merz presented a bill to Parliament that could have solved the "London problem." His frustrations were deep and his engineer's sense of economy and rationality was sharply offended when the bill failed to pass. Merz needed a bill from Parliament permitting a power company to lay distribution lines under London's streets. Those in Parliament who denied him this were of varied persuasion, but the primary objections came from members who wanted to protect the jurisdiction of the numerous vestries, parishes, and other political units that made up greater London. Some of these political, or local, authorities owned gaslighting plants. Others had invested heavily in direct-current or simple alternating-current systems. The extent of these light and power systems was contiguous with the political boundaries of the local authorities. Whether privately or government owned, the electric systems were for the local authority easy to regulate and, if privately owned, simple to socialize if the technical and political boundaries coincided. Merz's plan, though rational and attractive to the engineering mentality, violated the jealously guarded political power of long-standing authority. Political and technological power were at odds and, in London, the older of the two prevailed. It was about this time that Lloyd George told the young Merz that electrical matters were political, not technical.[7]

The Regional Opportunity

The appearance of steam and water turbines in power plants after 1900 brought unprecedented concentration of power, technical solutions, and technological problems. Station engineers and managers originally introduced turbines as a technical solution to the space problem caused by the monstrously large reciprocating engines. Steam turbines also offered operating economies. The largely unforeseen consequence of the introduction of the turbine was a near insatiable quest for load to fulfill the economy-of-scale potential of a large efficiently loaded turbine. The turbines were, in effect, supply seeking demand. The large turbine stations acted like technical solutions generating technological problems. If one is seeking illustrations of technological determinism, then the forces put into play by the coming of the turbine will serve. The turbine proved, for instance, to be an engine-driving organizational growth. Before World War I, the geographical extent of most utilities did not provide a load matching the supply potential of turbine central stations. Utilities had obtained their franchises, or concessions, during an era of small-area direct-current distribution and reciprocating steam engine generation. With the organizational structure outmoded by technical change, utilities began to reach out for new territory and new customers by means of high-voltage transmission lines.

The introduction of the turbine with the associated high-voltage transmission lines generated unprecedented capital requirements for the electric supply industry. Since the era of Edison's urban direct-current stations, the cost of distribution lines about matched the cost of generating equipment. Now the cost of a high-voltage transmission system was layered upon low-voltage distribution costs. The transmission system not only included the overland steel transmission towers and copper or aluminum lines with costly insulation, but also a complex array of lightning protection apparatus, circuit breakers, switches, transformers, synchronous generators, and motor-generator sets. The capital demands of the utilities during the 1920s, when many regional systems were under construction, exceeded that of the railroads during their decades of most rapid expansion.

Because advocates of soft or appropriate energy technology today sharply criticize the large turbine power station with a farflung transmission system, the contrasting arguments of those who built them are interesting. Earlier advocates of large regional systems argued that the unit cost of generation was an inverse function of the size of the turbines and generators. They appealed to the long-accepted propositions that capital cost per unit of power generated (kilowatt hour) decreased with the physical size and capacity of the machines. The reason for this was a matter of geometry because heat and magnetic field losses increased with surface area and the ratio of surface area to capacity decreased with the physical size of the turbine or generator. Others pointed out that large regional systems improved the load factor. The load factor was a measure of the utilization of the power station over a given period of time—day, month, or year. Since capital charges were paid on the full capacity, a higher load factor meant a reduction of unit cost. The large-area, regional station presented the opportunity for higher load factor because a more diverse load was likely to be found in a larger area of supply and a more diverse load usually meant a better load factor because the various loads did not peak at the same time of day, month, or year.

But, as observed, most utilities early in this century did not have the organizational structure and extent to encompass the envisaged regional systems nor the financial power to raise the necessary capital. Several critical problems of a nontechnical nature prevented the fulfillment of the technological potential. The situation then was not unlike today when advocates of various responses to the energy problem lament the economic, organizational, and other factors frustrating technical change. About the turn of the century, financiers, managers, and engineers invented a solution to the technological problem—the electric-utility holding company. Persons with long memories and economic historians need not be told that not everyone considered the holding company a solution. For many advocates of public ownership and for many foes of monopoly, the holding company was a monstrous problem, not a solution.

Were Holding Companies the Answer? Holding companies, however, proliferated, especially in the decade after World War I. As early as 1924 holding companies dominated the privately owned electric-utility industry. They controlled two-thirds of the generating capacity of the electric-utility industry. Seven holding company groups controlled 40 percent and other holding companies 25 percent. Holding company structure and financial policy were ingenious inventions. The engineers and managers of the large utility-holding companies elaborated upon the earlier holding companies in railroads and other industries in highly interesting and refined ways. Two of the larger, Electric Bond and Share (organized by General Electric) and Stone and Webster (organized by two Massachusetts Institute of Technology engineers who for years sat at the same desk and used a common signature) followed the standard financial practice of issuing their marketable common stock using as equity the unmarketable common stock of small utilities. They went beyond this however, to provide rationalizing management and engineering services for the utilities they combined. The holding companies also physically linked by transmission lines the utilities held, providing, thereby, the sought after load area for the giant turbines.

The onslaught of the Depression, however, wrecked the frail pyramidal structure of several holding companies, including Samuel Insull's, and raised doubts about the advantages of all of them. The mounting doubts culminated in 1935 with the passage of the Public Utility Holding Company Act, which sharply curtailed and reversed the spread of the holding company. The anticipated rise of super holding companies that would preside over regional systems covering the entire Northeast or Far West, even the nation, was not to be. Instead the private utilities and the federal government found themselves locked in a long-term struggle for regional system control and ownership, a struggle manifest in the fight over TVA and in frustrated efforts to create other TVAs. America, like London—and England—decades earlier saw technical implications contradicted by what Lloyd George labeled "politics."

Mixed Ownership in Germany. In Germany, the same opportunity to build regional systems presented itself. The German response in several cases differed from the prevailing response in the United States. The focus here will be upon the policy pursued by one of Germany's two largest utilities, the Rheinisch-Westfälisches Elektrizitätswerk (RWE) which supplied the highly industrialized Ruhr region. To better comprehend RWE policy, one needs to review some relevant highlights of the firm's history. After 1902 the RWE was headed by Hugo Stinnes, one of the world's most influential industrial leaders. Under Stinnes the RWE declared as its goal the supply of power at the lowest possible prices in the largest possible quantities. The RWE, like large American utilities, strove to increase gross output by lowering the income per unit output. Hugo Stinnes enunciated this policy in the Ruhr about the same time Samuel Insull was articulating it in Chicago. Before World War I, Stinnes, advised by his chief engineer Bernhard Goldenberg, used the technology of large-scale turbine generation and transmission to interconnect power plants using various forms of energy and supplying consumers with various kinds of loads. The goal was, to use present-day terminology, an "economic mix" of power generation and a high load and diversity factor. These technical measures were accompanied by organizational innovations including merger, joint stock ownership, interlocking boards of directors, and other combining business forms.

Throughout its history the RWE has also confronted the tension between public and private ownership. Before the war, the management, government administrators, bankers, and other concerned persons found ingenious ways of reducing tension and resolving conflict. An outstanding example of this German problem solving involves the RWE and three utilities owned by local government. The confrontation was resolved during the period 1906-1908 by negotiations. The affair is complicated, but instructive. In 1906 the RWE reached out toward the city of Dortmund and its environs. The RWE built a power station to supply the district and obtained supply concessions. It offered to purchase the city-owned utility in Dortmund. The city refused and, supported by Berlin banks and Walther Rathenau of the German General Electric Company (AEG), enlarged its own plant. The expansionistic policy of the RWE was further frustrated by other local government utilities in the region, one based at the city of Bochum and the other at Hagen. In this situation, Stinnes and his associates and the Dortmund utility agreed to form a new utility to take over the RWE power plant near Dortmund and the concessions that the RWE had recently acquired there. Not only would Dortmund and the RWE own stock in the new company, but also the communal utility at Bochum. The complex settlement provided for the new mixed-ownership utility to supply electricity to the communal utility at Hagen as well. By 1920 the RWE had a strong tradition of cooperative arrangements with local communities. The RWE itself became a mixed ownership enterprise with about half of its shares in the hands of local and state government and the other half in the hands of private investors. Because local government in the Ruhr region was strongly influenced by industrial leaders in coal, iron, steel, and chemicals negotiations were facilitated.[8]

I have suggested—and this may seem unusual coming from a historian specializing in technology—that the organizational, financial, and political problems have proved more intractable in the history of electricity supply than the technical ones. The policy implication of this chapter should now be obvious. If technical change has been retarded or furthered by nontechnical factors—and in many cases it certainly has—then policy emphasis today should be shifted from the technical level to the deeper structural, or technological, level. Policy makers should ignore lamentations about irrationality, social lag, politics and so on and realize that these will not go away after a dose of sweet reason, but are the heart of the matter. Lloyd George was right when he told young Merz that solving London's electrical supply was a political question.

CAN HINDSIGHT TEACH US TECHNOLOGICAL FORESIGHT?

Leslie Hannah, Director of the Business History Unit at the London School of Economics, observed that two basic points in Hughes' presentation seemed incontestable. First, technical historians have too often concentrated their attention on individual components and failed to develop as technological historians capable of treating complete systems. Second, it is dangerously simpleminded to brush away "politics" or "economics" as peripheral to problems, although engineers often do so, thereby forsaking their opportunity of solving them. Hannah felt that his own survey of electric-utility systems in Britain confirmed Hughes' analysis of the British experience.[9] There was no British Edison, except perhaps for Charles Merz, who was allowed to have his way only in the Northeast, where he built up an efficient, integrated, investor-owned electric utility. Merz was no laggard in associating with inventive engineers, but none of them put together a technical package as effectively as Edison, one capable of working as a technological system, in Hughes' terminology.

The illogicality of the British electric utility system in the first decades of this century drew much comment from engineers, but given its political origins it is appropriate that it was eventually the politicians who solved the problem. And in solving it late in the 1920s the politicians in Britain had one overwhelming advantage. By then they could look abroad to America and Germany and see the technological future they wished to create. The result was that in 1927 the Conservative government of Stanley Baldwin established a state-owned enterprise, the Central Electricity Board (CEB). The CEB was charged with building a national "grid" to interconnect all the major power stations, build new ones, and direct their operations. The lag behind America in efficiency and output was quickly made up as the benefits of this technical, organizational, and political innovation were manifested. Output was concentrated on the most efficient stations and new stations of optimal size were built. Generation costs (including capital costs) fell to one-third of their former level within a dozen years. Between 1929 and 1935, the output of electricity by public utilities in Britain increased by 70 percent, compared with only 13 percent in Germany.[10] This late breakthrough in Britain suggests that uncertainty is perhaps the greatest barrier to rational thinking on the planning of technological systems. Things seem to be easier when a "follower" country can use overseas experience as a blueprint and translate it to meet national political and economic needs. Much the same points are made more generally in relation to "follower" nations by economic historians. The recent advantages of rapidly growing countries like Germany or Japan compared with slowly growing ones like the United Kingdom and the United States derive not only from the higher research and development costs carried by the latter, but also from the vision of the technological future which the follower nations have been able to see, particularly in the United States, thus avoiding the mistakes of organizational transition.

Lessons about handling technological systems can, then, clearly be learned by studying other countries at different levels of technological development at one point in time, when the available technology may be taken as given. But one might well be skeptical about the potential for learning lessons from history in a constantly changing technological world. The temptation for the historian to impose a determinism on his data is strong: he knows what happened and does not have to construct a predictive model which can be proven wrong by events.

Yet there is nothing more difficult than learning from history—or, to put it another way, from experience, for history is nothing more than collective, recorded experience. Bad luck in the past can make a man overcautious in the future. Reading about the difficulty of developing, say, a nuclear technology in the past and the fortunes that have almost been lost in it, may unreasonably deter businessmen from taking risks in the future, when from a social point of view, economic risks are entirely justifiable if the whole basket of risky projects *on average* turns out better than the average of safer investments. A man who has experience of using consultants on power station design will refuse to believe that design work within the utility can be efficient and vice versa. As one surveys the diversity of historical (and individual) experience, it is difficult to resist the conclusion that hindsight and experience are as frequently used to close off desirable avenues of development as to open them up.

Arguably, the real lesson to be learned from history is the need for creative individuals who can escape from the bonds imposed by past human experience. Such individuals may develop a mental agility and an intellectual adventurousness from the study of the diversity of history—perhaps this is what Professor Hughes is really advocating—but if they try to deduce oversimple lessons from history, they may well find themselves imposing unnecessary intellectual blinkers on their imagination. Did men like Edison and Bell learn from history or did they determine to overcome it? And more generally in business, have the attempts to learn from the past too often themselves been out-of-date? Alfred Chandler, for example, has pointed out that the innovative managers who have transformed the organizational shape of American industry would have got nowhere if they had followed the contemporary teachings on organization of the Harvard Business School or Massachusetts Institute of Technology.[11]

Those who prefer a more scientific route to the same conclusion may refer to the work of experimental psychologists, such as Baruch Fischoff. Historical hindsight, far from liberating the subject's spirit and disciplining his judgment, tends to perpetuate a false certainty and a futile search for scapegoats for past failures. Fischoff's appeal for greater humility in our lessons from history is an appropriate one. Let us not go so far as to echo the wag who opined that the only lesson to learn from history is that we do not learn any. But it is difficult and greatly taxes the historian's expertise. *Good* history, toward which Hughes gave pointers, has a great deal to teach us.

GERMAN UTILITIES: THE PAST AS A BURDEN ON THE FUTURE

Peter Weingart of the University of Bielefeld in West Germany concurred that the history of technology is not adequate as a source for understanding social systems. General historians ignore technology all too often, but historians of technology overestimate its importance. The predominance in histories of technical matters of hardware isolated from the circumstances of its creation, socioeconomic or even political and cultural impact acceptance, and diffusion is not just the all-too-narrow subject matter of a relatively small and unimportant academic discipline. Rather, this technocratic perspective prevails among the vast majority of planners, engineers, administrators, and policy makers. To underline the systems character of technological innovation, as Hughes has done, is thus of the utmost importance not just for the history of technology but also beyond it.

As anyone familiar with the notion of systems in the social sciences will readily admit, there is one analytical problem implied, namely the points of reference in respect to which a system is conceptualized. Not surprisingly, in the history of technology one deals with technological systems in which the technical component is the focus. The stories that we are being told legitimately center on the *technical innovation* and the nontechnical factors that affect its creation. Its acceptance or rejection are selected on the basis of implicit images of the context of technical innovation. This remains a purely academic problem as long as the particular choice of system remains part of a historical narrative. It does assume practical political relevance, however, when a *specific* notion of technological system is generalized beyond its historical context and thus is given a normative function. By sketching briefly the emergence and present functioning of the German utilities it can be shown how this system proved to be advantageous under certain historical circumstances and disadvantageous under present conditions. In other words, Hughes' tenet that the "present—and the future—does have a past" might be modified by saying that there is a past in the present which puts a burden on the future.

As in the United States, the beginnings of the production and distribution of electricity in Germany were marked by the initiative of private companies and entrepreneurs like Werner von Siemens and Emil Rathenau. The primary goal of these companies was to produce electrical installations for which they had to create a market and thus a demand. A condition was to obtain a large volume of capital and concessions from the communities. This was achieved by the creation of holding companies (and banks) which were then given a large part of the shares of existing or newly founded utility companies. It is considered a merit of the private electric companies to have paved the way for the high-voltage technology but some inbuilt flaws were already then apparent. The companies' goal to create demand for their products was achieved partly by consolidating monopolistic markets through binding the utilities by contract to buying their respective products. These and similar practices led to an early concentration in the electric industry and constrained competitions for areas of distribution. Toward the late 1890s the number of community-owned power stations increased rapidly as cities and townships had been convinced by the technical advantages of the new system and the handsome profits that could be made with it. They either bought the private companies or established their own. Thus private ownership was replaced, on the local level, by public ownership, but only so that private capital could develop new initiatives on the next level in building overland stations. Of 25 such stations built before 1900 in connection with experiments with new high-voltage technology, only one was a public foundation.[12]

The *Energiewirtschaftsgesetz* of 1935, the law regulating until today the relations between producers and consumers of electricity, left existing property rights largely untouched and excluded competition "for the sake of the general welfare." In place of market competition a system of partial price controls, supervision, and competitive constraints serves as a regulating mechanism. This quasi-monopolistic system has led to inefficiencies which only have become apparent and an issue of public debate as a result of the energy crisis and the search for alternative technical and organizational solutions. The structure of mixed ownership can today be considered an obstacle, both to the more efficient use of energy and the development of alternatives to nuclear energy. What is lacking is a coordination of the diverse public shares to achieve a comprehensive policy of electric power supply.

Why, in view of the still-growing importance of the mixed ownership company with a *de jure* public control, does it present increasing problems to a comprehensive and economically sound energy policy? Several reasons are apparent. First, public shares and voting rights are fragmented and connected to conflicting interests. Second, the regulatory and economic functions of the federal and state governments overlap and lead to conflicts of interest. The antitrust commission found that a separation of these functions is essential. Often they are both the responsibility of one and the same minister, and public involvement is so diverse that already the separation of these functions presents problems. The commission sees "the danger that the interests of the utilities furnished with market dominating power are identified with the public interest."[13] In a way, this verdict over the mixed ownership companies implies that the orientation to the realization of profits and capital rent prevails over the orientation to providing optimal public service and to realizing technical and scientific opportunities.

The focus of attention shifted from the technical to political and economic elements within the technological system of electric power. The time span has been brought to the present in order to face the challenge of linking the history of technology to policy making. This level of general analysis supports Hughes' claim: there are indeed many analogies between the situation at the end of the nineteenth century and energy problems today. Lloyd George's dictum that electrical matters are political, not technical, is more adequate than ever. Thus, indeed, the present does have a past. But if the German response to the challenges of building regional systems seemed to be a solution to the "political" constraints, the historical analogy catches up with it, too. Mixed ownership and the implicit fragmentation of public control and regulation allowed the enormous increase in turbine capacity, the realization of economies of scale, and the consequent concentration in the electric industry. However, this sytem which has evolved over nearly a century and is now firmly institutionalized, today seems to have outgrown itself. On a more concrete level, in evaluating mixed ownership today, the past does not have a future! Purely technical solutions to energy problems facing policy makers and administrators are likely to become a burden on the future. The vision implied in this is to keep the present open for the future by what might be called policies for "reflexive technologies." These would provide for the utilization of as much knowledge as possible on the social, economic, and political implications of new technical systems *before* such systems are implemented. Reflexive technologies, then, are those realized in social systems rather than as technical elements in isolation.

NOTES

[1] This chapter is based upon research done for a history of electric light and power systems in the United States, the United Kingdom, and Germany.

[2] Philip H. Abelson, "Five Years of Energy Paralysis," *Science* 201 (September 1, 1978): 775.

[3] National Resources Committee, *Technological Trends and National Policy, Including the Social Implications of New Inventions*, Report of the Subcommittee on Technology, June 1937.

[4] Amory B. Lovins, "Energy Strategy: The Road Not Taken?" *Foreign Affairs* 55 (October 1976): 74.

[5] Lynn White, Jr., "Technology Assessment from the Stance of a Medieval Historian," *American Historical Review* 79 (1974): 3.

[6] Robert Belfield, "The Niagara System: The Evolution of an Electric Power Complex at Niagara Falls, 1883-1896," *Proceedings of the Institute of Electronic and Electrical Engineers* 64 (September 1976): 1344-1350.

[7] John Rowland, *Progress in Power* (London: 1960), p. 47.

[8] Edmund Todd of the University of Pennsylvania assisted my research on the subject of the Rheinisch-Westfälisches Elektrizitätswerk.

[9] L. Hannah, *Electricity before Nationalisation. A Study of the Development of the Electric Supply Industry in Britain to 1948* (Baltimore: Johns Hopkins University Press, 1979).

[10] Political and Economic Planning, *Report on the Supply of Electricity in Great Britain* (1936), p. 124.

[11] A.D. Chandler, *Strategy and Structure: Chapters in the History of Industrial Enterprise* (Cambridge, Mass.: Massachusetts Institute of Technology Press, 1962); and *The Visible Hand: The Managerial Revolution in American Business* (Cambridge, Mass.: Harvard University Press, 1978).

[12] H. Gröner, *Die Ordnung der deutschen Elektrizitätswirtschaft* (Baden-Baden: 1975).

[13] "Strom: Es wird bewusst diskriminiert," *Der Spiegel* 12 (March 14, 1977): 97.

13
Human Population and Ecology
F. Kenneth Hare

The explosive growth of the human population and its effects on natural systems and exploitable resources were our primary concerns. Our charge, to "explore the changes resulting from scientific discovery and technical innovation in terms of the impact these changes have on the human enterprise, in the sense of where we have been, where we are now, and where we are going," made eminent sense to us in observing the centennial of Edison. Few people have done more to expand the technological achievements of mankind. We found ourselves drawn to look not so much at the hard sciences as to those dealing with human behavior, perceptions, health, and institutions. The subsequent discussion bore out this view.

The central social, economic, and political fact of our times, we argued, is the unprecedented growth in human population, which must continue for many decades at least, and which must, with equal certainty, come to an eventual end, since all exponential growth ultimately outstrips possibilities. Barring war and pestilence on an unprecedented scale, the human population must continue to rise from its present value of 4.1 billion to a much higher figure. Fertility rates in several of the large developed countries have fallen to, or even below, replacement level. Even in those countries, however, the age structure is such that total populations will in most cases stabilize only after several more decades of increase. In most developing countries, however, fertility rates are still well above replacement levels. Recent falls in some of these countries encourage the view that the future rise of population will be slowed. Meanwhile the earth's overall population increase remains near 1.8 percent per annum, with a doubling time of about 40 years. Overwhelmingly these new beings are injected into developing economies with low incomes and negligible savings.

The crude world data conceal the division of nations into two broad and inhomogeneous camps, those that have experienced the "demographic transition" to low fertility rates and those that have not, which are all poor. It has been widely assumed that the transition can be attained only by a rise in income—and that such a rise in income demands population controls. Japan has achieved this transition.

Demand for the world's resources of space, raw materials, and energy, and for the yield of its agricultural systems, clearly depends on the number of people and their income. The two influences described in the previous paragraph work in opposite directions. If numbers increase, demand increases; but if income increases as fertility falls, total demand may not increase. In practice, however, there is a long lag between the drop in fertility and stabilization of population. Large increases in demand must be expected for some decades after fertility rates fall. Most projections suggest that stabilization of world population will not occur until the third or fourth decade of the next century. Total population by that time should exceed ten billion, and may well be much larger. The actual figure depends critically on the date at which fertility rates fall to the replacement level in the most populous countries. It also depends on the absence of major wars, disease, and famine. Clearly this increase, which is in full swing, poses extreme challenges to the world economy, to social and political institutions, to the maintenance of an acceptable natural environment, to infrastructures—in fact to every aspect of human life.

ECOLOGICAL AND ENVIRONMENTAL IMPLICATIONS OF THE POPULATION EXPLOSION

Each person added to the world's population adds to the demand for fuel, food, clothing, shelter, water, and space—in differing amounts, according to his place of birth. Past attention has been devoted largely to the effects on ecosystems and environmental quality in industrial societies, especially to the impact of industrial residuals. Environmental economics, in particular, has been weighted toward such manageable problems, as has the search for technological solutions. The environmental movement has been heavily directed toward pollution abatement and the protection of amenity, as that term is understood in Western societies.

But the major population growth is in developing countries and involves an endless search for new land, new sources of water and fuel, and new infrastructures. India alone has increased the area of irrigated land by one-fifth since World War II and her use of fertilizers about fortyfold. These pressures have been intense in many parts of the tropical world, most of them less well off and less organized than modern India. Two broad environmental realms in particular have been under very heavy pressure: the desert margin and the tropical rain forest.

The desert margin—from true desert to the drier tropical savannahs and scrub forests—has included some of the most fecund areas of the earth. Along the northern Saharan edge, and in the dry belt of Brazil, local rates of annual population increase of four percent and higher (before emigration) have been common. In the Sahel, and in Ethiopia and Somalia, lower but still high rates have been typical. The result has been pressure for higher livestock numbers (often above long-term carrying capacity, at low technological levels) and for cultivation of marginal land. At the same time there has been a decay of traditional land use systems—for example the nomadic pastoralism of the Tuareg and Fulani of West Africa—and an introduction of Western medicine and veterinary science, which tend to augment pressure of both human and animal populations. The outcome has been acceleration of the desertification process under the impact of climatically normal drought—the loss of productivity of the local ecosystems through soil erosion, destruction of woody vegetation, and lowering of water tables.

The tropical rain forest, the world's primary store of carbon on land, is under similar attrition, in this case from two directions. The first is the impact of "slash and burn" shifting agriculture through much of tropical Asia and Africa, and parts of Latin America. The second is the systematic effort in Brazil and elsewhere to replace the tropical rain forest with crops or productive ranching land, under grass cover. A stable and highly productive example is the Atherton Tableland area of Queensland, Australia. The net effect seems to be a progressive areal reduction of high forest and its replacement by poor successional or grass-dominated landscapes that store little carbon.

These two examples touch upon the hottest environmental issue of the day, the build-up of carbon dioxide in the atmosphere. Rising carbon dioxide is an empirically proven fact. The reduction of carbon storage in forest biomass since Neolithic times is another. Much less well known is the impact of human population growth on the tropical forests. It has been conjectured that carbon storage is being reduced by one percent per year. Some workers have claimed that forest clearance is almost as important as fossil fuel consumption in contributing to the build-up of carbon dioxide in the atmosphere, which has major implications for the future of climate.

The carbon cycle just treated is symptomatic of present tendencies in ecological analysis. We are learning to treat many of the actions of mankind in terms of the biogeochemical and energy cycles of the planet as a whole. The shift of viewpoint has been from that of obvious, local ecological impact to that of the global assessment of how materials and energy are transferred between major reservoirs and how human interference alters these exchanges. This major shift in perspective not only affects ecology, but has major implications for the atmospheric sciences, oceanography, soil science, forestry, and the agricultural sciences. Moreover, effective management of the environment involves a close union of this enlarged ecology with economics, politics, and sociology. It is taking a population explosion to bring them closer together, one outcome we should welcome.

Confronted with this bleak analysis, the workshop decided to concentrate on a series of broad questions. What are the implications of population growth for future demand for resources of all kinds, and especially for energy? What are the ecological and environmental implications of population growth? Are any scientific developments in prospect that could ease the pressures? What is the likely impact of medical science on the population outlook?

THE ECONOMIC PERSPECTIVE

Ronald G. Ridker, Director of the Program on Population, Resources, and Development at Resources for the Future, addressed himself to the first question. There is, he said, supposed to be a vicious circle operating between population and economic growth. It stems in part from the observations that there is no country with a high income per capita that has not experienced a significant decline in fertility rates and no country with low per capita incomes that has achieved fertility rates near replacement levels. The problem is how to bring about a transition from one regime to the other when the presence of high fertility rates is one of the principal factors keeping economic growth rates low. While proposing to conclude with some comments about how this vicious circle might be broken, Ridker began with several generalizations, derived from recently completed Resources for the Future studies that will, hopefully, clarify the nature of the population-resource problems facing us.

First, no serious, global, long-term shortages of resources are likely to occur during the next half century solely as a result of population and economic growth. A few minerals of lesser importance could experience some price increases, but others, notably seabed minerals, could just as easily experience some declines in price. Liquid and gaseous fuels are no exception to this statement even though the world may well exhaust most major sources of petroleum and natural gas within this time frame, because we anticipate that such natural sources will be replaced by synthetics from other fossil fuels, and possibly also vegetation and organic wastes at prices that are not, in real terms, dramatically higher than current levels. This generalization stems from considerations of likely developments in both demand and supply for natural resources. On the demand side, we anticipate some slowdown in its rate of growth because of projections that both population and per capita output will slow down in the future.

Table 13.1. Selected Indicators of Population and Economic Growth, Standard Case

	Absolute Values			Annual Rates of Growth (%)		
	1972	2000	2025	1960-72	1972-2000	2000-2025
World						
Population (10^6)	3,799	6,446	9,364	2.2	1.9	1.5
GNP (10^9—1971 US$)	3,839	10,711	24,199	5.8	3.7	3.3
GNP per capita (1971 US$)	1,011	1,662	2,584	3.5	1.8	1.8
United States						
Population (10^6)	209	264	304	1.2	0.8	0.6
GNP (10^9—1971 US$)	1,126	2,429	4,243	3.9	2.8	2.3
GNP per capita (1971 US$)	5,395	9,185	13,968	2.7	1.9	1.7

The projections about population stem from recent analyses of data coming out of the World Fertility Survey, suggesting that population growth rates may have peaked in nearly all countries and that at least a few large countries are experiencing significant declines in fertility. On the basis of this evidence some projections of future population growth rates have been made for the world (see Table 13.1). While these projections are not nearly as favorable as one would like to see, they are encouraging and do indicate that something positive is going on. The projection that economic growth rates will also slow down starts with the United States, where such a slowdown is expected because of continued declines in work hours and labor force participation rates, shifts in the age, sex, and education characteristics of the labor force, changes in the composition of output (resulting in part from saturation in a number of material-intensive sectors), and transitional factors resulting from higher energy prices and environmental cleanup costs, conflicts over land use, and siting of operations considered dirty or dangerous. So far as the rest of the world is concerned, the slowdown is expected to result from the elimination of the technological gap that has allowed countries like Japan to grow rapidly by borrowing more advanced technologies, the emergence of the same set of problems facing the United States as other countries approach United States per capita income levels, and continued problems of adjusting to high energy prices, especially in developing countries.

On the supply side projections of "prospective reserves"—the amounts of resources likely to be added to current reserves on the basis of new discoveries and technological advances—are most important. Suffice it to say that these estimates are generally quite conservative. For example, since no published figures are available for a number of minerals in China, none were assumed to exist, seabed sources of minerals were not included, nonbauxitic sources of aluminum were not included, heavy oils and tar sands were not included. When these two sets of projections are matched, we derive the results indicated, that supply can keep up with the growth of demand without substantial price increases for most minerals and sources of energy over the next half century. This is not to say that efforts will not be made to increase prices, as OPEC has done, but that such price increases in real terms cannot be sustained in the long run. What will happen if this analysis is correct is that such increases will be forced back into line with other prices through inflation and exchange-rate adjustments, as has been happening in the case of petroleum. Such price and exchange-rate adjustments can be very disruptive, but the causal factor is not long term scarcity but efforts to move some prices out of line in the prevailing pattern.

Second, while we need not fear long-term global shortages, there are distinct, serious possibilities of shorter term global shortages and longer term, or chronic, regional shortages. Such problems are likely to be caused, as they always have been caused, by inequalities in the distribution of population relative to that of resources and production capacity plus changes in economic conditions that alter trade relations between countries. The most severe of such problems we can foresee in the future is likely to occur in the energy sector during the period when the world must learn to rely on fuels other than petroleum and natural gas. Just how difficult the transition period is likely to be will depend on the willingness of countries to cooperate and help out especially disadvantaged countries, as well as on the extent and quality of planning both within countries and internationally. Judging from recent experience, we cannot be optimistic that much will be done to ease these transitions until the problems are close to overwhelming us.

Third, even if there are sufficient resources, the world is likely to become a more dangerous place. The probabilities of serious disruptions in vital ecological or social systems is likely to grow. In part these growing risks stem from the fact that as depletion occurs, global production facilities are likely to become increasingly concentrated in fewer, more powerful hands (with some exceptions). But more important is the growth of nuclear power plants, with all the risks of accidents and diversion into weapons they entail, CO_2 emissions which if climatologists are correct can raise global temperature levels and alter circulation patterns in highly disruptive ways, toxic substances which could poison large land areas and the oceans for centuries, and similar factors that could eventually overload ecological and social systems.

The problem with such factors is that we know so little about them, we are, in consequence, largely stymied in deciding what to do. The buildup of CO_2 resulting from combustion of fossil fuels is a dramatic case in point. A number of scientists believe that an increase in CO_2 levels that raises global temperatures by 2°C would be highly damaging to the ecological systems on which human life is based. But the range of uncertainty is so great that responsible estimates of the date when such an event may occur extend from 1990 on the one hand to never (because of possible offsetting and feedback mechanisms) on the other—hardly a sound basis on which to decide what to do. The only fundamental way to reduce such risks and uncertainty is to push ahead with scientific research, to try to understand the forces governing our lives before they overwhelm us.

Fourth, a slowdown in population and economic growth will help to ease some of these problems, but not so fast and certainly not by so much as is frequently alleged. Most of these problems are the result of past growth which would be with us even if no future growth were to occur. What such a slowdown can do is to purchase time, resources, and additional options: time to overcome our ignorance and redress the mistakes of past growth, resources to implement solutions, and additional freedom of choice in deciding how we want to live in the future. These are potentially important advantages, but they must be utilized if they are to be turned into actualities.

How can such a slowdown in growth, in particular in growth of population, be brought about, especially if the vicious circle referred to at the outset is operating? Ridker indicated that the only way we can do so is through selective interventions in the economic growth process. That is, we must discover just what it is about economic growth that has led to declines in fertility in the past. Is it the associated improvements in education, health, declines in infant mortality, and improvements in life expectancy, increasing female labor force participation, and other factors already being studied? Or might there be some yet-to-be-discovered correlates of economic growth that seem to be doing the trick? Whatever they are, we must find them and then advance them into phase with general economic growth if the vicious circle is to be broken. To date there are many hypotheses but no solid evidence. It may be that in the end we shall find that more radical measures than have so far been contemplated will be necessary. Here again, our future depends upon advances in knowledge and scientific understanding, a slow and uncertain process. But there is no alternative to knowledge of our circumstances.

A FEEDBACK SOCIETY: VITAL PERCEPTIONS

John R. Vallentyne, of the Great Lakes Research Advisory Board, told the workshop that feedback is emerging as the dominant integrating principle in postindustrial civilization. It is the personal and essential key to guiding individual development and affords an adaptive approach to human behavior. It is applicable on all social levels ranging from local communities to nations. Focusing on relations rather than things, it is the principle responsible for the origin and evolution of ordered systems in the cosmos, including galaxies, the biosphere, life, and our own species. Feedback may appear to us to have many of the characteristics traditionally drawn upon by primitive religions in linking man to nature, and human individuals to an identifiable, if not personal, god. In an era of information overload it is the crucial criterion for distinguishing useful information from that which is trivial, irrelevant, or misleading. Opportunities for the detection and use of feedback in all its manifestations from social linkages to those between man and the rest of nature have expanded enormously with technological devices for the storage, reproduction, analysis, and dissemination of information. Most important, the return of negative feedback, which industrial civilization has so insistently postponed, could be so imminent and momentous in its implications that it will certainly attract our attention. This feedback will stem from both aspects of the "demophoric" explosion—overgrown populations and large-scale technologies (the term combines population and its resource demands).

The first phase in the transformation of industrial society to the "feedback society" occurred in the 1960s, although it was not generally recognized as such at the time. It began with the sudden recognition that pollution problems were worldwide and in some cases global in scale. Among the more notable perceptions were urban smog; water pollution extending from rivers, bays, and estuaries to large lakes and even oceans; the transfer of toxic chemicals such as insecticides, thalidomide, and mercury through food webs and ecosystems; the widespread dissemination of radiochemicals such as strontium-90 in the biosphere as a result of nuclear weapons testing; and the probable influence of carbon dioxide from the combustion of fossil fuels on the climate of the earth.

The reaction of industrial societies was to erase or reduce these feedback signals from the environment, which were telling us that something in our behavior was wrong. Although maladaptive, an important consequence and intermediate step in the evolution of the "feedback society" was the creation of "environment" as an all-encompassing umbrella over previously piecemeal measures pertaining separately to air, water, land, and living organisms. This led to significant improvements in the fit between man and environment as seen in the cessation of nuclear weapons testing, recolonization of the Thames River by fish, and new legislative and institutional measures to deal with a broad range of environmental problems. Yet, with the lack of a man-in-the-biosphere philosophy and a failure to view pollution as a feedback signal from the environment, the causes of the problems remained working in the background. New problems such as acid rain, resource depletion, and energy wastage began to appear.

The concept of "environment" is not enough. We will never be able to resolve our problems without the recognition that we were created and have evolved within a feedback system that is ultimately biospheric in its dimensions. Five perceptions are vital to the transformation from industrial society with its antipollution concept of environment to the "feedback society" with its focus on man-in-the-biosphere.

We must perceive that we are living within feedback systems and that feedback systems live within us—ecological in the former case, biochemical and physiological in the latter. The distinction between positive feedback (disruptive if continued ad infinitum) and negative feedback (stabilizing if continued ad infinitum) must be commonplace knowledge in the street. We must accept a certain amount of discomfort and pain as necessary in guiding human development and behavior; and that in most cases postponement of negative feedback is to the detriment of the individual and society. Much of the theory and philosophy of this is given by Sayre.[1]

Man is a demophoric species with internal (physiological) and external (technological) components of mass and metabolism. We are dinosaurs with detachable parts such as houses, cars, factories, governments, and breweries that, fortunately, we do not have to carry around with us. The basis of this perception undoubtedly goes back to the origin of tool-using primates; however, it was only recently given quantitative expression by Bryson and Ross, Margalef;[2] and Vallentyne.[3] On the basis that one ton of coal-equivalent per year corresponds in calories to the food consumption of 8.1 humans per year, Vallentyne calculated the demophoric energy consumption for various nations of the earth in 1973.

The fanciful concept that an external environment somehow or other surrounds us and yet is independent of us needs to be replaced with a perception of our bodies as components of living systems that include the life-support systems of air, water, food, and climate. We need to recognize that the reflection we see of ourselves in the mirror becomes nonexistent if deprived of nature's air for more than three minutes; water for more than a week or two; food for more than a month or two. We are dead if we breathe the air we exhale. We are parts of ecosystems which are in turn dependent on regenerative cycles of the biosphere.[4]

HUMAN HEALTH AND THE POPULATION PREDICAMENT

Patricia L. Rosenfield, although prevented from participating in the workshop by travel difficulties, contributed an answer to the third question. She is a member of the tropical disease and training program of the World Health Organization, although she responded in a personal capacity. The impact of medical sciences on human health, she wrote, leads us to ask two fundamental questions. Why invest in health and population planning? Whose health improvement and which populations are of concern? As investments these serve broader goals than economic development, improvements in health statistics, and lowered crude birth rates. Investments in these activities are investments in human welfare. An integrated approach with at least these elements, linked to development activities, may be the best way to improve human welfare. Despite the recent emphasis on integration of health, population, and development toward the broader objective of human welfare, many health and population researchers are balking at this approach. Issues of cost, scarce manpower, lack of organizational skills, and lack of effectiveness are raised as obstacles to integrating services in developing countries. Yet examples abound where integration of services with human welfare as the overriding objective has succeeded.

The second query—whose health improvements and whose population planning—relates to recognizing what the population at risk sees as its first priority. This question suggests the special predicament of the population planner. The decisions we are talking about concerning population planning are not decisions for us to make. They are individual "couple" decisions, as family planning workers often stress, and in addition they are powerful political issues at a national level.[5] The global interdependencies are legitimate concerns for the peoples of North America. Certainly increased population leads to increased resource use and abuse. But the countries facing continued high population growth today are not avoiding their problems because of a perverse desire to consume more and more resources. Countries in Central Africa, faced with rising infertility due to venereal disease and other problems, are not concerned with resource overconsumption due to population growth. In many countries human survival continues to be the overwhelming issue. For example, in North Africa, children under five account for 68 percent of all deaths.[6] Where infant mortality ranges from 116 per 1,000 live births (southeast Asia) to 173 per 1,000 live births (central Africa) compared to 15 deaths per 1,000 live births (North America), the desire to procreate is not only understandable, it is clearly the correct individual decision.

The countries experiencing high population growth are continuing to face high infant mortality, high childhood mortality, and comparatively low life expectancies. Food supplies continue to be limited in the face of unequal distribution not only of food resources but also of arable land and sufficient water to produce that food. Debilitating parasitic diseases are rampant despite decades of intensive efforts to control them. For the individual family without the benefits of reliable food supplies and appropriate health care, a large family is the correct decision. "Without children the family unit is unlikely to survive; with them there is significantly increased chance of survival and even the possibility of prosperity."[7] The role of medical sciences in the past has had somewhat limited impact on reducing mortality, let alone on reducing fertility.[8] Statistics gathered by the World Health Organization show that although adult mortality has improved in many regions, childhood and infant mortality are at far from desirable levels in the developing world.[9]

There is thus a major dilemma facing decision makers who are trying to improve the well-being of their populations. How should scarce financial resources be spent? The data on the effects of health investments on fertility and vice versa are unclear and often contradictory. Some studies claim that there is "little evidence for a strong causal link between health care and a decline in mortality, fertility, and completed family size."[10] In Thailand and Indonesia, for example, family planning programs rather than significant changes in economic or health conditions were apparently responsible for reductions in fertility: "The total fertility rate in Indonesia dropped by 29 percent between 1968 and 1975, without a significant improvement in the standard of living."[11]

Health and economic improvements alone may not affect fertility; the introduction of social improvements, however, apparently has succeeded in reducing fertility in a fairly short time period. This is exemplified in a dramatic way by fertility reductions which have been achieved in developing societies which have shown a commitment to social improvements, such as Sri Lanka, Kerala (in India), China, South Korea, Taiwan, and Cuba. This diverse list of countries, not all-inclusive, has some common elements, especially improvements in land distribution and literacy rates. Their political organizations differ considerably. Nonetheless, their common social improvements, not explicitly related to health and population, seem to have led to similar improvements in health and population statistics.

CONCLUSIONS AND DISCUSSION

Though the workshop's participants questioned the main speakers at some length, they did not challenge their broad conclusions. The workshop had difficulty reconciling the positions taken by Ridker and Vallentyne and noted that there is in general a wide gap between the outlooks of ecologists and economists when confronting the population-resources issue. Economists tend to be narrow, but precise; ecologists, to be much broader, but vague. Ridker, for example, argued for economic growth but *not* for unbridled population growth. Vallentyne was against growth when it brought us up against the limits of carrying capacity. The chairman expressed the view that their differences were semantic rather than real, differences of language and assumptions rather than of substance.

Some surprise was expressed at the statement that fertility rates had turned downward in most countries. This is a very recent development, and the reasons for it are not yet clear. If it continues, it means that the world may have reached the upper, stabilizing, arm of the S-curve of fertility. This has already showed up in the growth rate of global population. From a peak of 2.1 percent per year it has recently fallen closer to 1.8 percent. There is a long way to go, but it looks as if the ordinary people of most countries have persuaded themselves, or found acceptable means, to limit population growth. Hence the population "explosion," though still a dominant factor in world politics, no longer looks as unmanageable as it used to.

There was a general feeling that political actions by government were in any case unlikely to be very effective in this area, unless those actions were accompanied by incentives or disincentives, such as the Peoples' Republic of China program of penalties, recently announced, for even a second child in a family. Education offered more promise. It was argued that Western education had begun to create a more realistic attitude toward population and resource issues. But most people live in developing countries, where sheer survival is often the central fact of life. Education in such societies, for example among the few remaining Inuit hunters or among Chinese farmers, takes place in the fields or on the ice floes. The workshop was clearly sympathetic to the idea that marked changes in life-style and political assumptions would probably be needed, in industrialized and developing countries alike. The continuing availability of resources is obviously one area where science and technology must be counted on very heavily. Medical science will also play a significant part, but most constructively when it is part of an integrated approach in which population policy, health improvement, and economic development occur together. The principal conclusion reached, however, was that this huge issue—how to contain the explosive growth of human population and avoid excessive strains on the earth's ecological systems—was well beyond mere technological control. It would require wisdom, comprehension of nature, and of human nature, and a willingness to accept other people's burdens. Science and technology can help in the attainment of such goals, but they cannot deliver them to us.

NOTES

[1] K. Sayre, *Cybernetics and the Philosophy of Mind* (London: Routledge and Kegan Paul, 1976).

[2] R.A. Bryson and J.E. Ross, "On the Nature of Environmental Concern," University of Wisconsin Institute for Environmental Studies. *Studies* 3 (1972): 1-24.

[3] J.R. Vallentyne, "Today Is Yesterday's Tomorrow," (presidential address), *Verhandl. Internat. Verein. Limnol.* 20 (1978): 1-12.

[4] Great Lakes Research Advisory Board, *The Ecosystem Approach* (Windsor, Ontario: Great Lakes Office of the International Joint Commission, 1978).

[5] G. Zeidenstein, "Strategic Issues in Population," *Population and Development Review* 3 (1977).

[6] International Year of the Child, "Children and Health," *IYC/Ideas Forum Supplement* 17 (1979): 5.

[7] J. Ratcliffe, "Population Control versus Social Reorganization: The Emergent Paradigms," *International Journal of Health Services* 8 (1978): 561.

[8] T. Balakrishnan, "Effects of Child Mortality on Subsequent Fertility of Women in Some Rural and Semi-Urban Areas of Certain Latin American Countries," *Population Studies* 32 (1978): 135-145; T. McKeown, "Fertility, Mortality and Causes of Death: An Examination of Issues Related to the Modern Rise of Population," *Population Studies* 32 (1978): 535-542; A.R. Omran and C.C. Standley, eds., *Family Formation Patterns and Human Health* (Geneva: World Health Organization, 1976), p. 531.

[9] T. Dyson, "Levels, Trends, Differentials and Causes of Child Mortality—A Survey," *World Health Statistics Report* 30 (1977): 282-311; J. Vallin, "World Trends in Infant Mortality since 1950," *World Health Statistics* 29 (1976): 646-674.

[10] "The Big IF," *People* 6 (1979): 38; P. Kundstadter, "Child Mortality and Maternal Parity: Some Policy Implications," *International Family Planning Perspectives and Digest* 4 (1978): 75-85; Cassen, "Current Trends in Population Change and Their Causes," *Population and Development Review* 4(1978): 331-353; J. Ratcliffe, "Social Justice and the Demographic Transition: Lessons from India's Kerala State," *International Journal of Health Services* 8 (1978): 123-144.

[11] Kundstadter, "Child Mortality and Maternal Parity."

14
Medicine and Public Health
Merril Eisenbud

This centennial convocation was held during a period in which there is much criticism of technology. There are some people who wish for a return to the simple life of previous centuries. Science and industry are being maligned for polluting the earth and complicating our lives. It is certainly true that the detrimental effects of many forms of technological advance are everywhere to be seen, but on balance we are a fortunate generation that has at our command a variety of conveniences that could not have been produced in former times even with retinues of hundreds of slaves. Nowhere has this been so true as in the field of medicine and public health.

MODERN TECHNOLOGY AND HUMAN WELL-BEING

Our ability to employ the quality of life as a guide to decision making depends upon objective measures by which to judge it. I believe there are several. For example, what about life expectancy at birth? A person born in the United States in 1900 had an average life expectancy of only 47 years, compared to 71 years in 1970. Is this not one measure of quality of life? Our answer must take into account that this improvement has been due primarily to the great reduction that has taken place in infant mortality, and far less to improvement in adult health. Between 1900 and 1970, death during the first year of life for United States males was reduced by 87 percent. This is the main reason that about 20 years has been added for the average white male newborn. By contrast, a 40-year-old white male in 1970 had a life expectancy of 32 years, compared to 28 years in 1900, a difference of only 4 years. Nevertheless, the fact that infant mortality has been drastically reduced certainly adds to the quality of life for child and parent alike.

A second important criterion is the number of days of disability during the productive years of our lives. Comparative morbidity statistics are not available, but tables of vital statistics tell us that the death rate from tuberculosis in 1900 was 194 per 100,000 compared to 1.5 per 100,000 in 1970. Victims of this disease not only died prematurely, but suffered long periods of incapacity. The death rate from typhoid fever dropped in this century from 64 cases per 100,000 to zero, and influenza mortality has dropped from 200 to 3.7 per 100,000. Death from enteric disease has been reduced from 143 to 0.9 per 100,000. Even mortality from accidents has been reduced, from 73 to 47 deaths per 100,000 persons per year. In the past, for every case of enteric disease, influenza, or tuberculosis that terminated in death, or for every fatal accident, there were many days of dreary and often painful incapacity. Our generations are healthier, not only because we live longer but also because we suffer fewer days of disability during our productive years. Has not this improvement in the state of our health added to the quality of our life?

A third measure of the quality of life is the amount of leisure time available to the average person. A 70-hour work week was not uncommon in the last century and almost everyone worked six days a week. Regular vacations were enjoyed by only the well-to-do. In contrast, most people now enjoy a work week of five days and rarely more than 40 hours. The six-day week of 60 or 70 hours allowed hardly enough time for adequate sleep, household chores, and attention to the bare necessities of personal and family care. The shorter work week, together with labor-saving devices in the home, now makes it possible for most workers to enjoy 20 to 30 hours of leisure each week. These hours, which we now take for granted, are a luxury that has come to the average person only in this century. Our ample supply of leisure time is an important by-product of technological development, made possible by the enormous advance in per capita productivity that has taken place in the last century.

The index of national productivity per worker increased nearly sevenfold between 1879 and 1970. The fact that so many Americans are wealthy people by nineteenth-century standards is a by-product of that increased productivity and is a fourth criterion by which one can judge the quality of contemporary life. The average annual earnings of employees in manufacturing industries (in 1970 dollars) has increased almost seventeenfold since 1900, from $487 to $8,150. This is the reason why so many of our citizens are able to enjoy television, hi-fi, foreign travel, skiing, and numerous other wholesome (and some not-so-wholesome) leisure time activities. Technology has given us both the material means and the leisure time with which to undertake these activities.

Still another criterion of the quality of life are the conditions under which we work. We hear much today about the hazards of the workplace. Granted, there is still much to be done, but conditions in the workplace have improved to such an extent that for many years the worker has been exposed to less risk on the job than during the leisure hours. There are many reasons for this. Modern workplaces are better illuminated, the workers are better educated and better trained, and mechanization has eliminated many classical types of accidents (although it has also created some new ones). The National Safety Council reports that in 1912 about 20,000 workers' lives were lost because of accidents on the job. By 1970, when the workforce was more than twice as large and production more than nine times greater, there were about 14,000 deaths. The number of deaths from occupational accidents per unit of production has been reduced by 92 percent over a period of 60 years.

By the criteria enumerated above, the quality of life has greatly improved in this century and the improvement has come about very largely as a result of technological innovation. Tuberculosis came under control as a result of better nutrition, better housing, a shorter work week, and improvements in working conditions. Enteric diseases were prevented by improved methods of food preservation. Pesticides, antibiotics, and methods of immunization all played a role in control over other infectious disease. Technology, in one form or another, has been an underlying influence that made all these methods of disease control possible. Nevertheless, there are debits due to technology, and these must be recognized. Modern technology has given us the threat of modern war. Our dependence on the truck and automobile has resulted in urban sprawl, congestion, and noise. We use too much energy. And, ironically, the great reduction in mortality due to communicable disease has triggered a population explosion in the less developed countries of the world, which, if not controlled, will doom hundreds of millions of their citizens to starvation. We are seeking in this review of medicine and public health to evaluate the debits and credits that have been accumulated from a century of technological innovation. In these discussions, we must remember that health is more than merely the absence of disease, and should recall the official definition of the World Health Organization: "Health is a state of complete physical, mental, and social wellbeing and not merely the absence of disease or infirmity."

A CENTURY OF DIAGNOSTIC AIDS TO MEDICINE

Morris Shamos of the Technicon Corporation reviewed diagnostic instrumentation as a primary contribution of technology to the practice of medicine. Carl Wunderlich, a physician in nineteenth-century Leipzig, was the first to keep careful records of the temperature of the human body in both health and disease. It was he who showed that the essential feature in "fevers" was the variation of body temperature. The first clinical thermometers came into use in England around 1886 and were quickly adopted by the medical community.

The first significant contribution to chemical diagnosis was the discovery, in 1838, that the sweet-tasting substance in diabetic urine was identical with grape-sugar (glucose or dextrose). Suitable chemical tests for such sugar soon followed, particularly the famous quantitative method of von Fehling in 1848 and the equally famous Benedict reaction that followed. We now have still better tests for glucose in body fluids, but what is important to note is that despite this seemingly early beginning, clinical or biological chemistry is largely a twentieth-century development. Diagnostic laboratories were virtually unheard of prior to the 1920s. Since then, however, the field has grown at an unprecedented rate. True, some simple tests were known a hundred years earlier, such as for albumin in urine and the microscopic examination of the urinary sediment. But except for the sugar test referred to earlier, and a test for acetone in urine, it was not until the beginning of the twentieth century that clinical chemistry became a recognized diagnostic tool. In 1907 a chair in biological chemistry was created at Harvard for Professor Otto Folin, who was the first to undertake a detailed study of the chemistry of blood and urine, and who devised methods for determining the amounts of various substances in these fluids by means of color reactions.

Clinical testing has shown rapid growth. In the United States chemistry tests alone in the clinical laboratory totaled 800 million in 1970 and 1.6 billion in 1975; in fact their number has been doubling every five years since 1946 and there is every reason to expect it to continue to do so. Contrast this figure with the nine percent annual increase in the total cost of medical care. The laboratory test statistics are to be credited in large part to the role of automation.

Probably the most significant of the diagnostic tools derived from physics were x-ray devices, the first of several imaging techniques now employed extensively in medicine, creating the medical specialty of radiology as well as a whole new industry. Within weeks of Roentgen's first announcement physicians had begun to experiment with the new tool and it was only a matter of months before it was in diagnostic use in hospitals all over the world. In New York, several months after the discovery, Thomas Edison gave a public demonstration that is said to have delighted large crowds. As imaging technologies developed over the years it became possible to marry them to the computer. Computerized axial tomography produces images of cross sections of the body, so displayed as to afford true stereoscopic views. More than 1,000 of these costly devices are already in use within the United States, about 80 percent in large hospitals affiliated with medical schools.

PHARMACEUTICALS AND THE CHANGING CLIMATE FOR INNOVATION

Jerome Schnee of the Graduate School of Business Administration at Rutgers University reviewed the impact of government regulation on the pharmaceutical industry—historically one of the strongest foci for industrial innovation. Drug regulation by the federal government began in 1938 when the Pure Food and Drug Act of 1906 was greatly broadened as a result of the elixir sulfanilmide tragedy of 1937, which revealed the risks associated with new drugs and pointed up the need to test them for safety. Prior to the 1930s few drugs had been introduced and their regulation attracted little attention. Procedures for the premarketing clearance of new drugs were among the most important provisions of the 1938 legislation, which was intended to protect the public from untested and potentially harmful drugs.[1] In 1962 Senator Kefauver's Antitrust and Monopoly Subcommittee originated amendments which had two major objectives: closer control over the premarketing testing of new drugs and altering the criteria for approval to market them.

Closer control over clinical testing of new drugs was achieved by specifying the testing procedure a manufacturer must use to produce acceptable information for the evaluation of a new drug application (NDA). The sponsor of each new drug was required to submit a "Notice of Claimed Investigational Exemption for a New Drug" to the Food and Drug Administration (FDA) prior to human testing, the practical effect of which was to require comprehensive data on animal tests before the FDA would allow human trials. The 1938 act had prohibited marketing new drugs until the FDA determined that they were safe. The 1962 amendments required drug firms to provide evidence, according to statutory scientific criteria, that new drugs met the claims made by their manufacturers. Although the full effects of the 1962 legislation are still being debated, three major accomplishments may be cited. Ineffective new drugs no longer reach the marketplace, by virtue of the advance proof of efficacy required. Untested medicines are now relics of the past. And the accuracy of claims for prescription drugs is now subject to adjudication.

The 1962 legislation has had harmful effects, as may be seen in the declining number of new drug products introduced. Special concern has been expressed that potentially valuable drugs are being discarded because the animal testing requirements are so restrictive. The need for safeguards to prevent the loss of useful drugs becomes apparent from the comment that "if even one new drug of the stature of penicillin or digitalis has been unjustifiably banished to a company's back shelf because of excessively stringent animal requirements, that event will have harmed more people than have been affected by all the toxicity that has occurred in the history of modern drug development."[2] The major criticism of present FDA controls over drug development is that a drug's potential hazards receive more emphasis than its potential benefits. Wardell's study of the drug lag makes it clear that the marketing of new, valuable drugs in the United States has been delayed or abandoned because the emphasis of federal regulation is misplaced.

Declining Innovation

The rate of pharmaceutical innovation reached its peak between 1950 and 1959 as the accelerated growth of R & D spending resulted in an increasing flow of new products. During the 1950s more than 3,500 new products and dosage forms were introduced into the United States pharmaceutical market. New chemical entities are new products which achieve a considerable degree of chemical differentiation; about 12 percent of the new products introduced during the 1950s were new entities. During the 1960s both the volume of all new products and the number of new chemical entities declined. The average number of new products and new chemicals introduced between 1960 and 1969 was 160 and 22, respectively, as compared to 358 and 42 during the 1950s. The rate of innovation in the 1970s continues to be relatively low.[3] Pharmaceutical innovation has also become concentrated in larger firms.[4]

Hansen has estimated that the average cost and time of developing a new chemical entity during the 1963-1975 period has been $15 million and seven years. He also found that one out of eight new chemical entity development projects resulted in a marketed product.[5] Estimates for the pre-1962 period have been provided in the author's earlier study of the development portfolio of one large pharmaceutical firm. The average development cost of $1.1 million for 17 new chemical entities which reached the market is almost one-fifteenth of Hansen's estimate. Similarly, the average development time of two years and attrition rate of one in three were much lower during the period before 1962.[6]

Patterns of new drug introductions in the United States were compared to those in the United Kingdom by Wardell, who found that the number of new drugs exclusively available in the United Kingdom was four times the number exclusively available in the United States.[7] A related aspect of the international drug lag is the growing emphasis United States firms are giving to foreign research and development activities. Foreign R & D expenditures have historically constituted but a small part of United States drug firms' budgets for R & D. As the rate of growth in domestic R & D slowed in the 1970s, United States firms more than doubled their foreign R & D spending. The annual percentage increase in domestic R & D spending averaged ten percent between 1970 and 1977 compared with an annual foreign growth rate of 27 percent. As a result, the proportion of total United States R & D expenditures devoted to foreign activities has averaged 15 percent since 1972, about twice the earlier level.

Statistics on worldwide drug R & D expenditures make it clear that these outlays are growing much faster in other developed countries than in the United States. The rates of growth in pharmaceutical R & D for West Germany, the United Kingdom, Switzerland, Sweden, Japan, and France for the period 1969-1972 were at least twice those of the United States. This larger percentage rate of increase abroad is not due to a smaller expenditure base. In fact, the 1972 pharmaceutical R & D expenditures of the EEC were approximately one-third higher than those within the United States.[8]

Not all of the decline in innovation can be attributed to tighter regulation. The two decades following World War II saw the rapid development of significant new drugs, largely because of substantial advances in the biomedical sciences. Since then it has become increasingly difficult to improve upon existing drugs in many therapeutic areas and future advances are likely to be more difficult. Proponents of a "research depletion" hypothesis contend that the underlying stock of pharmaceutical research opportunities was depleted by the rate of innovation during the 1950s. Former RDA Commissioner Alexander Schmidt explains research depletion in his comment: "We have temporarily exhausted the exploitation of new concepts and tools. Truly dramatic new progress in medicine now waits on some basic innovation in molecular science, some breakthrough in our understanding of disease mechanisms, some new therapeutic concept or some new tool."[9] The most comprehensive and provocative analysis of the social impact of the new regulations was conducted by Peltzman, who concluded that consumers, as a group, incur substantial losses as a result of the 1962 amendments. He asserts that revised regulations have cut the number of new chemical entities introduced each year by half. This has meant social costs from illness and death. Moreover, the amendments have served to double the cost of new drug development and reduced competition. Resulting price increases have cost consumers hundreds of millions of dollars annually.[10]

In order to achieve technological success in the future, drug firms will need to emphasize "breakthrough-type" research. Traditionally, the discovery of most new drugs has been based on knowledge of structure-activity relationships. Compounds are modeled on an isolated natural product or patterned after structures that may already be known to give the required physiological response. Using this successful structure-activity approach, industry laboratories have discovered the majority of the new drugs marketed since 1950.[11] It is doubtful that this approach will continue to be as successful. As the quality of drug therapy has improved in many product categories it becomes increasingly difficult to develop superior products. And although the industry is already highly regulated, its environment may become even more controlled during the next ten years. Despite changes in the research situation, a more flexible, experimental, and responsive approach must be taken toward any future legislation. The impact of previous drug regulation must be taken into account in formulating policies for the future.

NOTES

[1] John B. Blake, ed., *Safeguarding the Public: Historical Aspects of Medicinal Drug Control* (Baltimore: Johns Hopkins University Press, 1970); Harvey F. Dowling, *Medicines for Man* (New York: Alfred A. Knopf, 1970).

[2] William M. Wardell and Louis Lasagna, *Regulation and Drug Development* (Washington, D.C.: American Enterprise Institute for Policy Research, 1975).

[3] Paul deHaen, *Non-Proprietary Name Index* (New York: Paul deHaen, Inc., 1977).

[4] Henry Grabowski, *Drug Regulation and Innovation* (Washington, D.C.: American Enterprise Institute for Policy Research, 1976).

[5] Ronald W. Hansen, "The Pharmaceutical Development Process: Estimates of Current Development Costs and Times and the Effects of Regulatory Changes," Working Paper GPB-77-10, University of Rochester: Center for Research in Government Policy and Business, July 1978.

[6] Jerome E. Schnee, "Development Costs: Determinants and Overruns," *Journal of Business* 45 (July 1972): 347-373.

[7] William M. Wardell, "Introduction of New Therapeutic Drugs in the United States and Great Britain: International Comparison," *Clinical Pharmacology and Therapeutics* 14 (1973): 773-790.

[8] Harold A. Clymer, "The Economic and Regulatory Climate—U.S. and Overseas Trends," in *Drug Development and Marketing*, ed., Robert Helms (Washington, D.C.: American Enterprise Institute for Public Policy Research, 1975), pp. 155-164.

[9] Alexander Schmidt, Paper presented at the American Cancer Society Writers' Seminar, St. Augustine, Florida, March 1974.

[10] Sam Peltzman, *Regulation of Pharmaceutical Innovation: The 1962 Amendments* (Washington, D.C.: American Enterprise Institute for Public Policy Research, 1974).

[11] David Schwartzman, *Innovation in the Pharmaceutical Industry* (Baltimore: Johns Hopkins University Press, 1976).

15
Urban Development
John P. Eberhard

Cities are not found objects or by-products of natural phenomena. They are designed and made by man. While they may be created on paper or by words in an imaginary sense, they become real places (real estate) to be used by real people (citizens) only by the design and production of components and systems by the purposeful use of tools and technique, which is, of course, technology. It can be primitive or sophisticated. It can be capital intensive or labor intensive. It can be a haphazard collection of techniques or a systematically arranged process of production. Whether cities are conceived and built well or poorly, they still result from a set of technologies. Whether the human settlement is as simple as the Gambian village of Joffure, the village celebrated by Alex Haley in *Roots*, or as complex as Manhattan, its buildings, movement systems, sanitary systems, energy systems, and communication systems are designed and made with technologies which humans have developed and know how to use. Technology works no miraculous cures and by itself affords only the potential for a better life. Lewis Mumford told us long ago in *Technics and Civilization* that technology "exists as an element in human culture and it promises well or ill as the social groups that exploit it promise well or ill. The machine itself makes no demands and holds out no promises; it is the human spirit that makes demands and keeps promises."[1] It is this harnessing of techniques to human aspirations that seems to me to offer new potential for the character and quality of our urban places. I propose a basis for optimism about the revitalization of urban settings for the human family.

SCIENCE, TECHNOLOGY, AND ADVANCED URBAN SYSTEMS

Necessity is the mother of invention but technology forms the context for the innovation. For example, the need to be able to see when it is dark stimulated a long series of inventions. For most of history the devices used to produce light were simple applications of the physics of fire—fireplaces, candles, and oil lamps. The introduction of the light bulb was not possible without a technological sophistication that made it possible to manufacture it and situate it in a delivery system for electric power. This was only one invention-based urban innovation a century ago. Others included steel structural members that made tall buildings possible, elevators that made them usable, indoor plumbing, central heating, telephones, automobiles, and subways. This collection of inventions was essentially developed between 1880 and 1892; since that time, no invention of similar magnitude has displaced any of these solutions to human needs. If we examine the conditions that contributed to this remarkable twelve years of inventions, we find that there were new necessities and *new technologies*. The new necessities were centered in the rapid growth of cities occasioned by the movement of business and industry from craft-oriented "cottage industries" in villages and small shops to large manufacturing plants requiring the support facilities and employable populations of cities. Land costs rose dramatically, density increased, sanitary conditions spawned epidemics, and horse-drawn vehicles polluted the streets. These new necessities were met by a remarkable collection of fertile minds motivated by a spirit of invention and entrepreneurship rampant in the post-Civil War United States. So many inventions were produced in the last quarter of the nineteenth century that Congress seriously considered closing the patent office because "surely we have invented everything the human mind can possibly imagine."

The technological context was a major contributor to making it possible for many of those inventions to reach the marketplace as usable innovations. We had discovered fossil fuel energy and fortunes were being made in mining and marketing fuels. Major technological advances emerged as ways were found to harness these fossil energies to large tools that displaced the more limited efforts of water-powered devices or animal-operated machinery. Not only were the tools larger but the use of electrical controls made techniques more sophisticated. This technological breakthrough, combined with the social invention of the corporate form of doing business, was given by Mumford a well-earned name, the "Industrial Revolution."

The major cities of the world have grown at different rates since the turn of the century, but everywhere the major elements of the cities are based on the nineteenth-century inventions mentioned above and everywhere the technologies used to install them are little changed since the Industrial Revolution. Tools for digging and lifting have become larger and more sophisticated, concrete techniques have often displaced steel construction, lighter metals such as aluminum have become common, but by and large designing and building cities is a labor intensive process with unsophisticated tools.

Most current observers of the urban condition can recite a long list of problems: growth is too rapid in Mexico City, São Paulo is polluted, New York is dirty and broke. Washington is unsafe. Los Angeles is suffering from the high cost and pollution of a transportation system based on the automobile, San Francisco is thirsty, and Buffalo is cold. But problems are also opportunities. They are symptomatic of unfilled human necessities. We can add to these unmet necessities some additional pressures at work on the cities of the world: the shortage and increasing cost of fossil fuels, an end to the cheap availability of certain metals, strong governmental interests in increasing employment, and unremitting increases in population.

In these circumstances, we find a set of preconditions for a new burst of inventions and innovations that will displace those nineteenth-century urban innovations. We may very likely experience another 10 to 20 years of urban inventions that now will be conceived within a technological potential much different than the Industrial Revolution. The tools and techniques of our most advanced technologies are based on electronic data processes that merge design concepts and production tools with sophisticated methods of management. Production technologies of the Industrial Revolution are being replaced by production processes that can truly produce "diversity within unity." The invention of new materials and composites is being generated by the performance potential of the tools for forming and shaping them. Management techniques which have emerged in large-scale space and weapons systems programs have made it possible to invent some solutions on purpose rather than leave all new ideas to the random flash of genius. Finally, we have learned to use ignorance as a resource—to discover what it is that we don't know rather than depend on trial and error.

We have also developed during and since World War II a new form of social invention which combines public funds to cover risks in research and development too great for private initiative while still enlisting the production ingenuity and knowhow of private industry. We know how to use advanced technology to make the resulting systems responsive to performance criteria based on human needs. I would not wish to be taken as a pollyanna where technology is concerned. Technology is a scare word or even a pejorative term for many people because it tends to be linked with concepts of the use of techniques that are beyond the control of human intervention or management—or so it seems to many people. I would suggest that what advanced technology challenges is the human intellect and the paradigms for understanding and creating that most of us have grown to trust. We have always been uncomfortable with powerful tools we do not think we understand or which we must rely on others to turn to our use. As problems become more complex, so do the opportunities. What is one generation's limit becomes the next generation's basic vocabulary. Our cities, and their associated systems, have reached dimensions too complex for nineteenth-century solutions. The mind of man has made the leap to surround new complexities before, and we can do it again. It won't be easy or comfortable, but it is possible, challenging, and creative in the most basic sense.

SMALL BUSINESS AND SUCCESS IN INNOVATION

Dr. Leo S. Packer formerly served as counselor for science and technology on the United States Mission to OECD in Paris. As an engineer with 20 years' experience in industrial R & D management, he has come to be dissatisfied with the response of government and economic institutions to innovation. The United States share of world exports of manufactured goods has declined from 25 percent to 18 percent since 1958; our increase in output per hour in manufacturing since 1950 has been one-eighth that of Japan, one-quarter that of West Germany, and less than that of the United Kingdom. Public equity issues of small companies now number only 30 or so each year, down from more than 500 in 1969.

According to Dr. Packer, the economy today faces challenges of achieving a better environment, renewing blighted inner cities, developing alternative energy sources, and conserving energy and resources. Small innovative enterprises, whether oriented toward needed services or toward new technology, can play important roles in rebuilding urban communities.

With innovation, new opportunities and options become available for new technically oriented small businesses in revitalizing communities. These include new types of building designs, construction and renovation; design, installation, and maintenance of cost-effective decentralized solar devices; urban farming and small-scale food processing; specialized computer-based education and training centers; new technology centers for appliance and automobile maintenance and repair; better localized health care; and private delivery of welfare services. The list of needed social services that can be addressed by innovative small business is almost endless. Widespread participation in small enterprises gives control to residents of the inner city and provides long-absent economic opportunities and incentives for self-reliance and success. Most important, urban revitalization that is based on diverse profitable enterprise rather than bureaucratic, centralized public programs will provide the community a means of being self-sufficient and responsive to changing needs. Yet we also know that the Carter Administration's widely advertised multibillion-dollar urban program announced in early 1978 has produced little related legislation and still less favorable impact on our troubled cities.

The solution must be found in job creation—especially skilled jobs in the private sector. Innovation plays a key role, for employment gains come primarily from new businesses and products, founded on new concepts and technologies. Small businesses have an impressive record of creating new jobs through new technology. A recent study for the Technology Advisory Board of the United States Department of Commerce found that between 1969 and 1974 employment increased at an annual rate of 0.6 percent in a sample of large, mature companies, 4.3 percent in established innovative companies, and 40.7 percent in young, high-technology companies. At the same time employment in companies characterized as lacking in innovative activities actually declined. Industrial societies must be reluctant to protect established industry from the displacement of old products, although there are many pressures upon them to do just that, and instead encourage innovation, which affords the least inflationary way to increase growth in overall demand.

UNCONVENTIONAL TECHNOLOGY FOR THIRD WORLD CITIES

Jorge Wilheim, former secretary for economy and planning in the state of São Paolo, Brazil, described the strong challenges to innovation presented by the process of urbanization in the Third World. He emphasized that urbanization in most developing countries follows a trend and rapid rate of growth that leaves them unable to keep up a decent quality of life for the vast majority of their citizens. In Brazil, domestic migration is still responsible for 50 percent of the annual growth of São Paolo, a metropolitan region of 11 million (one-tenth of the nation) growing at a rate of 4.3 percent per year. What underlies this trend?

Domestic migration is a typical symptom of underdevelopment. The modernizing sectors of the economy, those linked to industrialized countries, are so privileged that they exert undue influence on other regions. Their privileges are based on their complexity, which requires geographic concentration in a circle of causes and effects that widens the gap between the regions in which the modern sector is concentrated and the rest of the country. The result is conflict between a very modern, highly concentrated, fairly sophisticated urban life in the Brazilian south; and outside it, a vast geographic periphery motivated to migrate inward. In such a context, urbanization takes on a specific meaning.

Since the Middle Ages to live in a city has meant *freedom:* freedom to move around, to pick a job, to break free from family and community social constraints, and to meet and mate freely. Today this freedom is limited by socioeconomic constraints. Opportunities are unequal. Poverty generates social disruption. Violence and repression severely limit the freedom expected in urban life. What seems most poignant in the urbanization saga of the Third World is the frustration arising from the contrast between actual opportunities and the expectations aroused by the vast array of advertised products constantly offered to the population. We live under a tyranny of desire that creates, even in the poorest migrants' minds, psychological craving for objects. It is this consumer's style of life that migrants identify with "modern life" in cities. In the context of underdevelopment urban life is a highly dynamic mixture of migration, constant social mobility, expectations of modernity, and frustration. Infrastructures and services are lacking everywhere, further undermining the quality of life.

How can this prospect be avoided? On the one hand we must move toward decentralization in budgeting and decision making. On the other we must provide effective technical procedures for equipment and infrastructure, primarily by employing what are called alternative technologies. Peter Drucker says technology means "man at work." To a filmmaker (Trnka) technology is the son of human laziness, of a compulsion toward least effort. Many others consider it a way to mix capital and human resources with machinery. When we make statements such as "the most efficient" or "the best solution," we should remember that different meanings for technology can multiply the meaning of such simple statements. If we expect technology "to fulfill human needs" we should be prepared to explain whose basic needs we are talking about.

At present most of the technology employed in the Third World is developed in Europe and the United States. It comes to us mainly through the import of products. When industrialization is substituted for the import process the original technology usually continues and machinery is imported for it. At present only four percent of the world's R & D is conducted in developing countries and even that small amount is mostly consecrated to technology copying that of the industrialized countries. Scientific and technical education is based mainly on European and North American patterns. We are offered only temperate solutions to tropical problems. The technology transfer process is thus one more means of creating the small, privileged, modernized sector that prevents broad-based development.

Who can doubt that this process is heightening in its effects? Beyond the 1980s some 400 transnational corporations will produce about 40 percent of the world's goods. This may be the structure of modernity but it should not be regarded as the only avenue of development. Development should direct economic growth toward the creation of jobs, standards enhancing the quality of life, and social equity. Technology is being patterned to modernity, not to development in this sense. Consider that sewage technology is derived from the ancient Roman "cloaca maxima." In industrialized countries more than 90 percent of homes are linked to networks of this kind. In the tropics only five or ten percent of homes are linked to central sewage networks. Is there an alternative to Roman technology for this region? If there is, it must be created, it cannot simply be transferred.

Technological alternatives that might be sought for the Third World include sewage treatment for small groups of homes, capable of generating biogas for cooking, human resources systems employing schools to follow up and advance students' employability, decentralized traffic systems including special corridors for experimental forms of public transportation, and neighborhood approaches to public health. These might be twentieth-century innovations to add to the nineteenth-century technology on which most cities are still based. The time for innovation may be more limited than we are accustomed to thinking.

In underdeveloped parts of the world cities are growing most rapidly. In ten years, eight out of ten of the world's largest cities will be in the Third World. They must have new equipment and a new infrastructure, serving basic needs in new ways achieved by lowering resistance to innovation. These cities constitute a laboratory for major innovations. Even transnational corporations could make good use of this experimental field so long as their contributions are socially appropriate. Financial institutions and international organizations should take a much deeper interest in urban innovations in developing countries. These countries must at the same time rid themselves of technical prejudices and academic preconceptions derived from nineteenth-century technological patterns in the temperate regions of the world. In order to rediscover the basic needs to be served they should find new ways of motivating and listening to people, partly by decentralizing power and decision making. Technological innovation in urban areas, as has been said of electric power elsewhere in this volume for an earlier period in history, is thus a political issue, but the human prospect should not elude us because of that fact.

ENERGY ASPECTS OF URBANISM IN THE THIRD WORLD

Samual M. Berman, senior scientist at the Lawrence Berkeley Laboratory, said that in comparison to the cities of the industrialized countries of the world, the urban metropolitan areas of the developing countries have a much higher likelihood of experiencing severely critical shortages of food, resources, and energy. This crisis situation is mostly due to the unprecedented increase in the population of these urban areas (urbanization). The promise of a better life brought about by employment and educational opportunities as well as vastly increased human interaction all provide an ever-present attraction for the poor rural population to migrate to the urban areas with perceived opportunities. This migration, superimposed upon an urban birth rate that is still high in many societies, leads to increases in the population of the developing urban regions from three to ten percent per year.

While the rural dweller most often uses little marketed energy, and relies upon noncommercial forms for basic energy requirements, the situation changes markedly upon the transition to an urban environment. Cities are engines for development; therefore, increases may be expected in commercial energy use, principally for transportation, electric power, manufactured goods, and services.

The coupling of the large increases in urban populations with the sharp increase in commercial energy use means that the developing countries often expect 400 percent to 600 percent increases in energy requirements by the end of this century, although per capita the rise is still modest. This situation is in sharp contrast to the energy needs of the industrialized countries where 50 percent increases in energy are considered reasonable. Herein indeed lies an energy crisis with critical portent for the entire world.

A further exacerbation of the energy problem arises in considering the amount of energy exported from the cities of developing countries to the rest of the world. Should the effect be as large as has emerged from analyses of Hong Kong and South Korea, then the requirements for energy demand in the future will be notably greater than the quantities we, or any predecessors, have projected. Extrapolations of demand to 1995 could reach as much as one-third of the world total, which in light of higher prices seems unlikely to eventuate. It is more reasonable to expect that these developing metropolises will find ways to diminish their energy use. The innovations and responses to population growth that will develop over the next several decades should cause unprecedented social, institutional, and environmental changes which could profoundly affect the developing countries and their urban settlements.

In order to illuminate the various components of urbanism and energy in the developing regions and how they might respond to this energy crisis, we have attempted to portray how typical megalopolitan regions will develop and organize themselves during the next 15-20 years (pointing to the period when present policy shifts would have maximum impact). From the pictures that emerge here some insights can be given where technical, organizational, substitutional and other techniques may be introduced that can ameliorate the impending energy crisis. In particular, resource conservation must become the primary principle of design for these urban centers. Energy use plans should be drawn up and enforced by appropriate institutions. Innovative housing and transportation systems must be developed. Indicators of efficiency in the use of energy must be found, and made prominent in the public context. Needless to say, these are novel considerations for foreign policy analysts and diplomats but they must now become primary for the entire world.

NOTE

[1] Lewis Mumford, *Technics and Civilization* (New York: Harcourt, Brace and World, 1934), p. 6.

16
Food and Agriculture
René Dumont

Growth in world agricultural production accelerated in the eighteenth century in England, rising by 40 percent between 1730 and 1780, creating a material basis for the Industrial Revolution. Growth speeded up further in the nineteenth century with the spread of chemical fertilizers and advances in genetics. It gained even more momentum with the internal combustion engine, for the tractor replaced animal energy with greater power and the use of nonrenewable fossil energy did away with the need to produce the local, agricultural, and renewable energy that had been obtained through feeding draught animals. It might be thought, therefore, that an era of universal abundance had arrived and that the specter of famine, which has always haunted mankind, was about to be banished forever. Nothing of the kind occurred!

From Agricultural Plenty to the Hunger of the Third World

Affluence has been increasingly ill-shared and inequalities in income have multiplied since the Industrial Revolution. Paul Bairoch considers that the gap in average individual incomes between the major nations of the globe was no more than 1 to 1.5 or 1.8 in 1700; today it is thought to be 1 to 40. The world economy is organized by those who are at the center of the capitalist system and run it to their own advantage. In North America cereal consumption amounts to one ton per inhabitant per year, almost entirely consumed by animals, and meat consumption has doubled since the last war. At the other extreme, the majority of the rural population of the Indian subcontinent do not have enough to eat and fall below the Indian government's poverty line. From the Sahel to Ethiopia, from the Andes to the Brazilian Northeast, from Java to the Philippines—*the hunger zone*—the disgrace of our era, is not growing smaller: far from it, the number of the hungry and the semihungry is on the increase. Malnutrition and undernourishment are killing off an extra five million people a year, 12,000 a day!

We Are Pillaging the Third World

For a long time, Europe dominated and organized the planet to its own advantage, more particularly by developing slavery and colonization. Genocide was the fate of the American Indians and of the Africans through the slave trade. The result of all this was that, in the tropics, agricultural export products, initially sugar and spices, were favored at the expense of food products for the local populations. In many cases, Europe stopped producing textiles and oil seeds, and went to find them in the tropics, where it created "beautiful" plantations of sugar cane, oil palm, Hevea rubber, tea, coffee and, of course, spices. It went to find jute, sisal, and cotton and stopped cultivating flax and hemp. Peanuts, copra, and palm oil replaced the colza, rape seed, oil poppy, walnut, and olive of the temperate countries. It is these export crops, which monopolized the most fertile land and the strongest workers, that are mainly responsible for the hunger now prevailing in the Third World. It might have been thought that this situation would quickly improve when the former colonies became independent, but nothing of the sort happened.

Neocolonialism Is Making the Situation Worse

Colonialism, by introducing the money tax and the import of products manufactured in the developed world, had already aggravated the situation of peasant life in the tropics by levying higher taxes on agricultural production, which increased very little, if at all, in unfavorable natural and socioeconomic environments. Moreover, by reducing mortality it triggered a great population explosion. The elite installed in power through independence, often by the former colonial power itself (e.g., French-speaking Africa), have sought to adopt our civilization, which is so absurd and so wasteful and is characterized by the private automobile. In order to live at this high level in such poor countries, they have had to corner the largest share of external "aid" for the benefit of the privileged urban minorities in power and exploit their peasants through taxes, trading profits, and, above all, duties on export crops. They therefore continue to favor export crops, always at the expense of food crops. Hunger therefore continues to increase, yet other threats to our future are looming on the horizon.

We Are Destroying Our Forests, from Amazonia to Thailand

Admittedly, for the past 30 years, Thailand has been developing its agricultural production at the very remarkable annual growth rate of five percent. Its exports amount to more than two billion dollars a year (cassava, maize, and sugar in addition to the traditional rice) and its population, which is growing at the rapid rate of three percent a year, is less badly fed than the majority of the Southeast Asian countries. Yet it destroys more than a million hectares of forests a year, exploiting the fertile wealth accumulated by centuries of forest land as if it were a mine. At the present rate of destruction, Thailand's forests will be gone before the end of this century. The Ivory Coast is destroying them even faster; within ten years they will be completely cleared. Brazil has already destroyed one-half of its forests and is attacking Amazonia to establish very unproductive ranches, designed to glut the rich countries with meat, even though they already eat too much of it.

Again, the forests on the slopes of the Himalayas, in India and Nepal, are fast being destroyed, either to serve as fuel (Calcutta) or to be worked by poor Indian peasants who are deprived of croplands by inadequate agrarian reform. This makes the rate of flow of the rivers even more irregular and aggravates floods and droughts, when the forest humus which used to retain a considerable part of the monsoon rains disappears. The moisture-laden winds on the Bight of Benin no longer meet up with the well-stored humidity of forests which are receding from the shore and the Sahel receives less rain. Have we, like sorcerers' apprentices, begun to effect what might be an irreversible change in our climates?

We Are Killing the Soil: Erosion and Deserts Are Taking Over

In the sandy soil (only five percent clay) in the north of Senegal, humus used to provide the "binder" which kept in the fine matter, and long fallow periods, when no crops were planted, allowed the spontaneous growth of vegetation which, once it had taken root, restocked the soil with humus. With the increase in population and, above all, the spread of peanut cultivation (in Senegal it accounts for a larger area than the nourishing cereals millet and sorghum), the fallow recedes and the humus diminishes rapidly. The wind erosion (think back to April 1934 in the center of the United States)—the harmattan blowing in the desert from January to March—then uproots all the soil's most valuable fine matter (loam, clay, humus, and fine sand). All that remains is a skeleton of coarse sand, like dunes along the seashore, fields which are swept away by the wind and no longer have any capacity to retain water or nutrients. In short, the desert is taking over, as it does through overgrazing of steppes, improper cutting of thorny acacias in dry years, and deforestation for fuel.

This advance of the deserts is such a threat to the planet that the United Nations Environment Program convened a conference on the subject at Nairobi, Kenya, in the summer of 1977. One of the conclusions reached there was that the peasant should be capable of restoring organic matter to the soil, which presupposes that he should no longer be robbed of the fruit of his labor by the developed countries (unequal trade) and by the privileged urban minorities in power in his own country. It also presupposes that the poor countries should be helped to fight against deserts; a contribution from the rich countries, estimated at $400 million a year, was recognized as necessary. The oil-producing countries quickly agreed, but we, the "old rich" did not. The world spends more than $400 billion a year on armaments but cannot envisage spending one-thousandth of that sum in order to guarantee the future of the human race.

From the Volga basin to Algeria (where 50,000 hectares of land are lost every year) and the Andean sierras, water erosion on the slopes is growing worse. The South American landowners monopolize the plains and underutilize them as pasture on very large ranches. They oblige the peasants, who are driven back onto the steep mountain slopes, to work them and in this way accelerate the erosion. The plains should be worked and the slopes given over to grasslands and forest so as to safeguard the heritage of mankind. But the big landowners are not in the least worried. What is needed is agrarian reform, admittedly a difficult undertaking to carry through successfully, but one that nevertheless is absolutely essential.

The United States Agro-Food System Is Squandering Scarce Resources

The high productivity resulting from the advanced application of the science and technology in North American farming, with only four percent of the active population engaged in agriculture (and perhaps one percent by the end of the century), has long been praised. It is now recognized that the United States' agro-food system is far too wasteful of energy and metals, which will soon be in short supply unless something is done about it very quickly. Less than six percent (and soon five percent) of the world population thus wastes more than one-third of the world's energy and principal metals. We, the participants in this symposium, use up an average of 500 times more energy and metals than does the poor peasant of the Sahel, Bengal, or the Ecuadorian sierra. By what right? By the right of the richest and strongest. However, this is precisely the agro-food system that, particularly through multinational corporations, we (the scientists, technicians, and economists) are now spreading throughout the world, especially the Third World, under the very improper name of *development*.

As these countries are too poor to adopt it, the system is available only to the privileged minority in power and, in order to "benefit" from it, they squander external "aid" (a very inadequate quid pro quo for the pillaging of the Third World) and exploit "their" peasants. This means that, although the gap between developed and underdeveloped (or even developing) countries continues to grow, in poor countries the gap between town and country, between the elite and the peasants, continues to grow as well. The gap between Ouagadougou and the countryside of Upper Volta was estimated at five to one in 1960 but at ten to one after 1970. And there are so many poor people in the towns. The aid furnished to Upper Volta exceeds its total budget and nearly all of it finally ends up in the capital, a parasite town which more or less lives off aid. This means that aid increases the need for aid, creates even more dependence, and in no way leads the Sahel toward economic independence.

This System Is Leading Us to Our Doom

Many criticisms may be leveled against the conclusions of the Club of Rome, and particularly against its first essay "The Limits to Growth." But the essential conclusion remains, even though the time limits for the exhaustion of scarce resources envisaged in that essay may often be put back by a few decades; in our present life, with the twofold industrial and population "explosions," we are heading to our doom through exhaustion of our planet's scarce resources, air and water pollution that have become unsustainable, and perhaps an irreversible impairment of the climate. Military aircraft and the supersonic Concorde attack the ozone layer which protects us from ultraviolet rays, and the CIA tells us that we may be in for a general cooling down of the planet; others, on the contrary, fear that the increase in carbon dioxide will cause a general warming up, the melting of the ice caps, and a rise of 60 meters in the level of the seas. We do not know how to control the radioactive waste of nuclear power stations and soon any worker with a modicum of skill will be able to steal the plutonium from breeder reactors and manufacture a home-made atomic bomb.

Although we are growing richer and richer, there are fewer and fewer of us, whereas the poor, the often landless peasants of the tropical countries, along with the unemployed in the slums, are growing hungrier and hungrier but larger and larger in number. Thus, given the continuation of present trends and our political and economic system of increasing domination and inequality, to believe in a peaceful future for our planet seems to me to be the worst kind of pipe dream. We must envisage more reasonable utopias, reducing injustices, and safeguarding the future of humanity, if we still believe in it.

The rise in the consumption of agricultural food products is the result, in roughly equal proportions, of the overconsumption of meat by the rich and of the increase in the number of the poor, although in their case their needs are never fulfilled. Admittedly, we advise them to bring a halt to their population explosion, which would indeed reduce their distress, but first of all they must be allowed to develop all of their production capabilities, all of their "hidden productive forces," as the Chinese say. To begin with, productive work should be found for peasants, instead of leaving them semiemployed, semistarved, and crowded into slums on the fringes of cities. The rich North American consumes 500 times more energy than the Afro-Asian peasant; the European, 200 times more. It is therefore essential to begin by cutting down on the wastefulness of the rich.

From a Profit-Making to a Resource-Saving Economic Structure

The claims advanced for the capitalist system that it is highly efficient certainly seem true when it is compared with Soviet agriculture! Yet the frantic competition that is its hallmark has already brought us two world wars and many more minor conflicts. And it seems particularly effective in what Pierre Jalée aptly describes as the pillaging of the Third World. Surely the privileged, those of us present here, already have enough. The United States imported 400 million tons of oil in 1978. By 1983 the figure may well be 500 million! Hence inequalities multiply even more rapidly than differences in food production.

The most intelligent capitalists, like my friend Aurelio Peccei, leader of the Club of Rome, tell us that the misemployment of science and technology by the capitalist system may lead us within the next century to certain doom if we continue our insane industrial expansion, our frenzied urbanization, our mindless squandering of resources, our frantic buildup of arms, our unsustainable pollution. The present economic system must, if it is to endure, accumulate capital uncontrollably. So far it has not been able to envisage the survival of mankind and the reduction of inequalities, the protection of nature, and proper food for the hungry. It is, quite simply, leading us to our doom. Consequently, we have to change it and this is not an easy task, since the failure of systems of the Soviet type, though not of the Chinese type, is acknowledged.

The aim is survival and hence a more equal distribution of income. A start could be made by canceling the debts of the poorest countries. This would be a realistic step, since they will not be paying them off, and a moral step, since these debts are in fact the outcome of underpayment for their raw materials and labor and overpayment for our manufactured goods. Survival calls for the saving, the judicious economizing, of all the scarce resources of our planet and, as a start, energy, metals, and water. In a market economy, these savings could be encouraged by following the suggestion of the British ecologists, i.e., very high taxation on energy, on metals, and on water and air pollution. In this way, the economy would be steered more and more toward the saving of the scarce resources of our planet. In addition to the royalties of the producer country, oil would be subject to an international tax for the development of food crops in the poor countries and a higher national tax that would encourage everybody to take trains, to drive less, and to build or convert housing with better insulation, which would require less fuel for heating.

Dearer energy would make for speedier utilization of the energy of the sun, the winds, and the tides, and in agriculture there would be more natural air drying of corn in corn cribs. Agriculture could produce its own energy through fermentation of plant waste, as in the case of manure gas, the biogas of China and India, which would call for an enormous amount of new research, for science managed in another way and aimed at economizing, at survival, and at the least possible inequality. Science is, in the final analysis, a tool that the powers have so far misused and abused. Therefore, we need to "set the powers right."

Politics, Always in Command, Needs to Be Shifted toward Survival

A good many scientists claim that they are "apolitical." This position is, politically speaking, very clear, for it is one in favor of the status quo, in other words, of maintaining the system that is hastening us to our doom. The scientist is responsible for the use made of his research and the atomic scientists who condemn nuclear weapons have realized this. Once scientists understand that the direction now taken by technology is leading us to doom, their *duty* is to take an active part in the debate. In the tropics we have not yet developed a system for continuous cultivation of the soil without fallow land, one that guarantees continued fertility and is within the reach of the peasants. The peasants still have hoes, whereas computers are creating ever more opportunities for us. But we are continually perfecting weapons, and in the Netherlands geneticists are trying to find new colors for their flowers.

It is not a question of condemning science—quite the contrary. Indeed, the ecologists' concerns for respect for the planet, the concerns of people for the hunger in the Third World, compel us to embark on an enormous amount of new research, but, more than anything else, on a different approach to the application of research. To do this, we need to single out the privileged, to denounce them as the seeds of the destruction of mankind and, from among the poorest people, regroup new political forces whose program would be the enduring survival of the human race.

Utopia: The Only Practical Alternative to Doom

The directions now taken in the applications of science and technology constantly make for greater inequalities among countries and among social groups. Some squander outrageously and do so more and more. Others, even greater in number, are without work and receive ever less food. The world is heading for disaster on "our little planet," as my friends Barbara Ward and René Dubos pointed out in 1972. The blame should not be laid on science, but on a disorganized application of science that affords extravagant privileges for the powerful. Once it is acknowledged that such policies are leading us to our doom, we have to seek a social, economic, and political framework that is capable of survival, and the ecologists are starting to work one out. But a growing amount of new research will be needed. Hence, this position is not one that is hostile to science.

The most urgent and the most important thing is to aim at a new delegation of powers at all levels, with the priority set on survival and on the least possible inequality. Here are scientists with their backs to the wall, ready to take on their responsibilities and, with all their strength, enter into the field of politics so as to halt the race toward disaster. *The chances for the enduring survival of mankind are already seriously jeopardized.* If we do not work out and achieve, well before the end of this century, an economy that is based on saving the scarce resources of our planet, we are condemning our descendants, and it will happen so soon as to affect the youngest participants in this symposium. Are we going to be capable of dealing with the responsibilities that face us because of the disorganized application of our "scientific" work, which we have generally thought was being done for the good of humanity?

DISCUSSION

David Hansen, a member of the biology faculty of Pacific Lutheran University, summarized the points of view in the discussion as the dominant, disparate views of the world food situation. One view was that the gulf between the wealthy and the poor, and the concomitant high degree of malnutrition among the latter, has been accentuated by Western techniques of high-energy agriculture, which drive marginal farmers to urban areas because of the high economic inputs required, and high unemployment results. The other view was that self-sufficiency is ultimately attainable with the proper application of technology through irrigation, fertilizers, and appropriate genetic strains of crops.

Some common ground was found. Both agreed that poverty is the principal cause of malnutrition and that population growth and food production are closely correlated. Proper nutrition and a more secure existence would bring down the rate of population growth. But sharp differences arose as to the extent of the current problem and where ground is being gained or lost, how the gulf between wealthy and poor should be reduced, and whether in fact the poor may not actually be better off in areas with wider gaps when national wealth has increased overall. Opponents of the transfer of Western agricultural technology argued that in many cases such technology has led mainly to increased productivity in cash crops for export, whose price is controlled for the benefit of the few. This was claimed to have especially severe consequences for women.

Proponents of Western-based agricultural technology cited overall gains in per capita production and predicted that self-sufficiency would eventuate from broad application of proper methods. Opponents argued that major social reforms and land redistribution were needed, along with high taxes upon industrial consumption to curtail global imbalances in commodity and resource use. Both groups agreed that income yield for small and medium-sized farms had a vital part to play as an incentive toward self-sufficiency.

Emil M. Mrak, former chancellor of the University of California at Davis, offered an extensive commentary on agricultural technology in the workshop on medicine and public health. A summary of his comments is included in the discussion record here. Emotional attitudes toward DDT produced a rush to judgment against it, he charged, citing the coho salmon of the Great Lakes, the subject of a report that FDA tolerance standards for DDT had been exceeded. At once the Department of Health, Education, and Welfare recommended that DDT use be phased out. Then it was found that there had been an error in the analyses and that the material present in the fish was not DDT but polychlorinated biphenols from industrial wastes. Farmers have had to live with the chaos of resulting regulations.

Research is the only possible basis for hoping that pest control programs can adapt to changing circumstances and also help to control the losses that occur during the storage and transportation of food, which account for the loss of roughly half of agricultural output in some countries. In Japan, Professor Mitsuda has suggested drying rice to a low moisture level, sealing it in plastic bags, and storing them in lakes and tunnels, out of the reach of rodents and insects. The continued increase in energy use, whereby the equivalent of 80 gallons of gasoline goes into growing an acre of corn in the United States, can only be properly understood and perhaps corrected through the findings of research. Ronald Robie, director of the California Department of Water Resources, warns that the use of water for irrigation, which now accounts for 90 percent of consumption in areas such as the Sacramento drainage basin, is likely to be curtailed by future population growth.

Despite the urgency of research needs, private industry is cutting back on research and public attitudes seem to be forcing reductions in the government's agricultural research budget. Experiment station and extension service budgets are being cut. Most of the research conducted in industry might be called defensive rather than creative. How could the efficiency of photosynthesis be increased? How will we develop plants able to grow under adverse conditions? How will we improve basic methods of water use? Jonathan Swift wrote three centuries ago that if anyone could produce two ears of corn or two blades of grass where only one had grown before he would contribute more to the world than all politicians taken together, an observation whose truth still seems beyond challenge.

Dr. Walter Falcon, director of the Food Research Institute, Stanford University, observed that Professor Dumont had provided the workshop with a sobering description of the world food economy. He shared his concern about declining natural resource bases and his call for self-reliant agricultural growth in hungry countries. On other topics he sought to develop a different view, seeing the hunger problem primarily in terms of protein and calorie malnutrition, stemming in the poorest countries, such as Bangladesh, from extreme pressure on a limited land base, the effects of uncontrolled rivers, and low income. Claims about an exploitative export trade dealing in luxury foods are simply irrelevant to the major food problems in the country.

He differed from Professor Dumont on the impacts of technology derived from Western sources. New seeds, additional fertilizers, and improved irrigation systems were the basic reasons for the accelerated growth in agricultural production over the last decade. Thus Indian wheat production increased from 11 million tons in 1960-61 to 29 million tons in 1977-78. There is evidence to indicate that smaller farmers are gaining in absolute income terms as well.

Many of the future choices on agricultural technology will be difficult and the required investments in supporting infrastructure will have to be large. Agricultural research promises continued improvements similar to significant breakthroughs made on irrigated wheat and rice and to some extent on rain-fed wheat and corn. Creating appropriate agricultural technology, training the manpower to manage it, and taming flood-prone rivers will not come easily, quickly, or cheaply.

Finally, there are difficult decisions to be made in the years ahead with respect to energy. All countries could do a much better job of using organic wastes for agriculture. For the much longer run also, increased research may generate possibilities of nitrogen fixation among grasses, alternative energy sources, and new agricultural rotation methods. However, for perhaps the next 20 years, sustaining and increasing output on a global basis will be quite dependent upon heavier use of energy-intensive inputs such as fertilizer. Gearing up the research establishment to provide viable alternatives for the longer run that are less energy intensive is crucial. But just as important is assuring continuing availability of the roughly five percent of total energy use devoted to agricultural production worldwide.

UPPER VOLTA AND IRAN: SUBSISTENCE AND SOCIAL DESIGN

Thierry A. Brun, assisted by Sylvie Bonny, both of the National Institute of Health and Medical Research in Paris, offered specific analyses of agricultural patterns in Upper Volta and Iran to substantiate the theses presented by Professor Dumont. It is often suggested that poverty and underdevelopment result in part from the absence of adequate technological investments, that the application of science will readily solve the problems of underdevelopment. We fail to see that the wealth of some nations arises from the poverty of others and that science has been enlisted in this imbalance. After peanuts were introduced as a cash crop in Senegal, production mounted rapidly from 9,000 tons in 1870 to more than one million tons by 1976. Although this crop offered many possibilities, among which was the production of low-cost proteins for human consumption, the Europeans limited the processing of peanuts to the preparation of vegetable oil, while the peanut cake (or meal) was sent to Europe for use as animal feed.

As far back as 1911, the French inventor Rudolf Diesel had shown that his engine could function perfectly using peanut oil as fuel. As related by J. Albert in his book on vegetable oil in black Africa in 1941, Diesel designed his engine specifically for the use of locally available fuels in the mechanization of agriculture and handicrafts. The use of peanuts to convert solar energy into vegetable fuel for diesel engines might have proved more beneficial to tropical countries than the exclusive use of fossil fuels that has become customary. Also, peanuts have been shown to be good sources of protein for infant feeding. Robert Poivre of the National Institute of Health and Medical Research in Paris designed in 1957 a process to eliminate virtually all the potentially toxic components from peanut oil so that it could be used to prepare a protein-rich flour to supplement the inadequate cereal diet of the most vulnerable segments of the population—babies at the age of weaning, pregnant and lactating women, and the elderly. However, none of the governments of the Sahel and neither foreign investors nor international agencies ever decided to invest in the manufacture of such protein-rich mixtures despite the obvious need for them. While imported industrial milk products and baby foods were preferred by local ministries of commerce and health, local sources of protein were exported at low prices for the cattle-feed industry of Europe.

Because of soils, population density, and distance from the coast, Upper Volta was less favorable than Senegal for cash crops. It serves instead as a reservoir for cheap labor which is exported to neighboring countries such as the Ivory Coast. Programs of agricultural extension leave the subsistence sector of Upper Volta almost untouched. Despite almost a century of contact with European technologies, the subsistence sector has not benefited from any of the labor-saving, yield-increasing inputs of modernized agriculture. The large majority of farmers of the Sahel do not use chemical fertilizers, pesticides, improved seed, or mechanization.

The stagnation of the subsistence sector is characterized by the "vicious circle of hunger"—malnutrition causes high morbidity and limits work capacity, which in turn leads to low productivity and generates more malnutrition. The subsistence sector is not, however, isolated from the employment sector. It feeds the urban poor and also provides the market sector with young workers. In Upper

Volta it is estimated that the total number of migratory workers exceeds 20 percent of the active population. In some age groups (males 25-35 years of age) more than one-third of the people have left their villages to migrate abroad or to the capital. We have conducted an energy audit of subsistence agriculture in Upper Volta showing that cereal production has been declining so that increased imports of cereals are required to feed a formerly self-sufficient population. Moreover, we observed seasonal weight losses of 7.5 pounds for men and 2.5 pounds for women between January (the post-harvest period) and September (when the harvest begins). The expenditure of energy by males rises from an average of 10,200 Joules per day in January to 13,300 during the rainy season in July. Any significant drop in cereal production in this subsistence group, which makes up the majority of Upper Volta's population, exposes them to great risk of famine.

In Iran, despite the oil wealth available, the Ministry of Agriculture adopted policies to force the abandonment of traditional agricultural methods in what were called (and treated as) "marginal zones" while a selected number of areas were favored as "pilots" for "development." Most oil revenues dedicated to agriculture went for a few large-scale irrigation projects entrusted to large companies owned by associates of the Shah, which benefited from such government policies as tax exemption, water at subsidized rates, and special import rights for machinery and equipment. Tens of thousands of farmers were expelled from their land in order to make the large scale projects possible, and despite all the investment Iran's imports of food increased at a rate of 14 percent per year during the 1970s. For the DEZ dam, 47,000 families were expelled, leading to massive migrations to the cities affording a large supply of cheap labor for industrial projects. The resulting social imbalances do a great deal to explain the subsequent political outcome in Iran.

It is now just a century since Cornelius Walford provoked a debate at the Statistical Society of London in which most participants expressed their certainty that famine would shortly disappear from the world. Agricultural technology has been used to expand factory and cash crops, not to dispel famine or meet human needs.

WOMEN AND SOCIAL SYSTEMS: NEGLECTED ASPECTS OF AGRICULTURE

Susan George of the Transnational Institute, an element of the Institute for Policy Studies, called attention to the effects of technology on Third World rural women. Because the inventors of technology are generally men, and because women are not involved in its planning or implementation, the burdens it places upon them often go unrecognized. According to one United Nations Food and Agricultural Organization (FAO) study, women perform about half the tasks in a "traditional" village and up to two-thirds of the work in a "modernized" village. In one village chicken raising was introduced to increase volume, but the planners neglected to make provision for the extra water needed by the chickens. As women were the water carriers, the extra load fell on them. When villages do produce a surplus, women are often expected to do all the work required to move the crop to market. Women are sometimes unduly severely displaced by mechanization of agriculture, as in electrical hulling of rice in Indonesia, with depressing effects on their nutritional status.

When new technology makes a given activity more attractive, as in the introduction of simply pottery appliances in Ghana, women may be displaced by men. They are, in any event, often paid less than men for identical work, as in Sri Lanka's tea estates, where men make $14.00 per month, while the maximum pay for women is $11.40. Consider the effect that simple, fuel-conserving stoves could have in reducing the time women must devote to collecting dung or firewood. Mechanical handmills could replace the mortar and pestle for grinding grain. Better arrangements for fetching water would free women from much drudgery. Unfortunately, needs such as these are rarely taken into account by development planners, who are overwhelmingly male.

Is Overpopulation the Cause of Hunger?

While few would deny that rapidly growing populations in many Third World countries may intensify hunger, to assert that overpopulation "causes" hunger contributes little to solving the problem. Overpopulation is a meaningless term outside a socioeconomic context, as can be seen by comparing India's 172 people per square kilometer to Holland's 326 per square kilometer. The economic terms of access to land and the way in which the land is used seem to be more important than the total amount of cropland available. Asian countries commonly credited with solving their hunger problems have less arable land per person than India or Pakistan by factors of two (China) or five (Korea). According to the FAO, 2.5 percent of the world's landowners, with holdings above 100 hectares, control three-quarters of the world's cropland.

It has frequently been assumed that birth control technology will afford a solution to demographic problems. This is, again, to mistake the symptom, high population growth, for the cause. The Harvard "Khanna Study" and its follow-up indicate that even with intensive educational and extension measures and free distribution of contraceptives in a pilot project area in the Punjab, no dent could be made in birth rates as people have no real incentive to have fewer children. Their motivations for doing so are, in fact, quite rational in their perspective, for the poor in countries without social security or old-age pension need children. Parents count on their children to provide free labor and to help them survive in old age. Male-oriented cultures also exert strong pressures on women to produce the requisite number of boys. When demographers come to understand that Third

World people *are* planning their families inside the framework of constraints bearing upon them they will come to see hunger and rapidly growing populations as symptoms of the failure of political systems to afford people adequate means. When fewer children are needed and available land is more equitably used people will respond accordingly.

AGRICULTURE AND ECOLOGICAL INNOVATION

Pierre Spitz of the United Nations Research Institute for Social Development in Geneva observed that the general public perception of agricultural and food problems is often colored with ideas of natural obstacles and constraints, assuming once-for-all given conditions. But think how many dramatic changes have occurred in the world map of plants and animals since the sixteenth century. A static world picture dominated by constraints is inappropriate and misleading. Lewis Mumford described many of these changes in *The City in History* in 1961.

Research in North American agriculture has emphasized mechanization; in Europe, higher yields through fertilizers. In Third World countries, where the loss of a single harvest can mean death, the research undertaken must be aimed not only at obtaining a high yield but also at stabilizing it through good and bad years. The problem of stability is of vital importance for impoverished peasants. While a variety of millet able to weather drought was produced for trial at the International Crop Research Institute for the Semi-Arid Tropics in Hyderabad, Indian agricultural authorities were reluctant to approve its introduction because it didn't out-produce other varieties in good years. Another agricultural practice emphasizing stability is mixed cropping, which distributes risks of loss among several varieties. Professor S.K. Roy of the Indian Statistical Institute has shown that some traditional peasant practices such as planting two strains of rice in the same field are unexpectedly productive because of interactions between the different root systems. Diversified agricultural practices are also helpful in maintaining genetic diversity. Undue dependence on energy supplies also exposes agricultural systems to the risks of disruption if those supplies are interrupted. If the energy = civilization equation has to be revised, the corollary equation illiteracy = ignorance needs reconsideration as well. Underlying peasant know-how may be some very important latent principles whose development could be of great consequence for the Third World.

17
Democracy and Technology
Arthur Kantrowitz

The institutional processes of science subject the claims of theory to judgment and the facts of experiment to reconfirmation. The engineering professions maintain common frames of reference to assess the performance of technological systems. In the political realm disagreement is tolerated rather than overcome. Yet public programs increasingly have scientific components which can become infected by differences in attitude, purpose, or clientele, and disagreements of this kind cannot be reduced by the procedures customary to science. Yet the public is accustomed to appeal to scientific authority in the resolution of disputed matters, so the appearance of loud and prolonged controversies on matters involving science and technology has aroused intense concern lest the usefulness of science as the basis of policy judgment be undermined. Dr. Arthur Kantrowitz proposed in 1967 an "Institution for Scientific Judgment," since become known as the science court, to intensify consideration of publicly disputed claims potentially resolvable on scientific grounds. The science court has been proposed in a number of publications by Dr. Kantrowitz.[1] *He arranged a discussion of applications of the science court concept in a number of contexts, by contributors of the following papers. The session opened with Dr. Kantrowitz's oral presentation, summarizing the proposal and his rationale for its implementation. (Ed.)*

SCIENCE COURT PROCEDURES FOR COMPETING TECHNICAL PROPOSALS

James F. Decker of the Office of Fusion Energy, United States Department of Energy, described the application of the science court procedure to restricted technical issues not well known to the public, rather than to the major national questions for which it was initially proposed by Dr. Kantrowitz. His paper described an experiment in the use of science court procedures for resolving technical issues of a type often encountered in the management of technical programs. More specifically, in mission-oriented research and development programs there are often several potential solutions to a given technical problem. In order to apply limited funding to obtain the best solution with acceptable risk, the program manager must determine which solution is the most "promising." Even when no one approach stands out as clearly offering the best route to success a decision must still be made, based on a number of complex considerations. The magnetic fusion program provides an excellent example of this because many approaches to heating and confining the fusion fuel exist. Starting in 1977, Department of Energy's (DOE) Office of Fusion Energy, formerly the Division of Magnetic Fusion Energy in ERDA, evaluated a number of competing approaches to magnetic fusion using novel techniques as an aid to solving this management problem. The approach consisted of establishing well defined evaluation criteria with a quantitative scoring system and using an adversary or science court-like procedure for application of these criteria. For the particular application described here, the science court approach was found to be effective for reaching judgments for DOE management use in ways that may be of use to others.[2]

DOE's magnetic fusion program has as its primary objective the development of economical fusion power for commercial use. Fusion energy is one of three inexhaustible energy sources being developed by DOE. The other two are solar energy and the fission breeder reactor. Fusion is the process which produces energy in the sun and stars. Fusion energy is released when light elements are caused to fuse into heavier elements at very high temperatures (approximately 10^8 degrees Kelvin or higher). At these temperatures, the fusion fuel is in the fourth state of matter called a plasma in which the electrons are stripped from the atoms. In magnetic fusion, the hot fuel is confined and insulated from material walls by magnetic fields. At the present time the two main magnetic confinement concepts being supported in this program are the tokamak and magnetic mirror. Based upon the important favorable experimental results obtained in tokamaks, particularly the Princeton Large Torus and the Massachusetts Institute of Technology Alcator A, there is a high degree of confidence that a power-producing reactor can be built using a tokamak device. However, there exists the possibility that other magnetic confinement concepts might lead to more attractive fusion reactor systems than either a tokamak or a mirror. For this reason, DOE supports research on alternative concepts with the objective of performing experimental tests of critical physics issues as required to evaluate the viability of these concepts.

A number of alternative concepts at different stages of development are supported with varying levels of funding. Funding decisions which determined the pace of research on each of these concepts had been made in the past in response to individual proposals for support. This management approach led to a variety of technical approaches to plasma confinement but did not lead to a systematic identification of those concepts which are the most promising, which was required to apply the limited resources available to those concepts which offered the greatest potential advantages over our present mainline concepts. In 1977 DOE's Office of Fusion Energy evaluated the alternate fusion concepts it was supporting. Uniformity was sought by standardizing the criteria against which to evaluate all concepts and employing the same group of evaluators to apply those criteria to all of the alternate concepts. Three basic questions were asked. Assuming that all the required components can be built as specified, will the reactor operate and produce power? Can the individual components be developed? Assuming that the reactor can be built and that it will operate, is it desirable from an economic viewpoint? An evaluation grid was derived from these questions. A numerical scoring system was developed with the aid of many scientists and engineers from universities, national laboratories, the electric power industry, and utilities, which rated each of the areas in specific terms.

The Science Court Procedures

In order to obtain a uniform evaluation, all of the concepts were evaluated by the same evaluation panel, composed of appropriate experts. It was recognized at an early stage of developing this process that a critical evaluation of all 11 fusion concepts would require an inordinate investment of time and effort on the part of the evaluation panel if normal review procedures were used. Consequently an adversary procedure was adopted to streamline the evaluation while still retaining the standards of criticalness and thoroughness that were desired. Two other groups, called "advocates" and "critics" appeared before the evaluation panel. The advocates of each concept were the scientists and engineers working on it. The advocates were responsible for providing and defending scientific results and extrapolating from them as well as arguing for the attractiveness in technological terms of the reactor embodying them. A chief advocate was designated to act as their coordinator. The critics were chosen for their special expertise in the areas of physics or engineering more important to each concept. Their responsibilities were to ferret out crucial physics and technology questions. Advocates of one concept sometimes served as critics for the evaluation of another concept. Their work was coordinated by a chief critic.

The judgments of the evaluation panel were reached in a procedure we called "the conference," where advocates and critics debated the technical issues. Before it convened, the advocates supplied a written description of the concept, a description of the state of knowledge in physics on which it was based, with supporting references, a description of the reactor embodying the concept, and an evaluation of that concept employing the standard criteria. The critics examined the material provided by the advocates and provided their evaluation of the concept using the same criteria. The advocates and critics met in advance to determine points of agreement and disagreement and preclude tactics of surprise. The evaluation panel conducted the actual conference. The advocates were given about 40 minutes to advance each concept and the critics received equal time, followed by questions from panel members. The initial evaluation by the panel itself was undertaken in executive session immediately afterwards. The conference could be reconvened if one or two issues needed discussion in greater detail. After this a final evaluation of the concept was obtained by averaging the panel members' scores on each separate issue considered. The conference devoted about four hours to each concept.

Results of the Evaluation

The 11 alternate fusion concepts evaluated were the Elmo Bumpy Torus, slow liner, fast liner, Tormac, laser-heated solenoid, theta pinch, e-beam heated solenoid, reversed field pinch, multiple mirror, ion rings, and SURMAC. The detailed results of the evaluation are described in the technical report already cited. The results have been used by the Office of Fusion Energy as a basis for accelerating, redirecting, or terminating research as appropriate. The evaluation led to high scores for the EBT and liner concepts, which appear to be attractive in reactors that could employ technology near to the state of the art. Important unresolved technical issues were identified for further study. When a physics issue was involved in a judgment of weakness in a concept, near-term research was focused on that problem. Weaknesses in technology or reactor desirability served to suggest changes in the configuration of several reactor types. In other cases the concepts were rated so low that funding for them was terminated.

Within recognized limitations, this exercise was quite successful in meeting the objectives of providing a uniform, critical evaluation of alternate concepts of magnetic fusion. We were experimenting with criteria and procedures and intend to revise them in similar evaluations in the future, while also seeking to limit them to fewer concepts. The adversary procedure played an important part in the success of this evaluation, achieving a reduction in the time needed for technical review. Even though the issues confronted were primarily technical and did not involve social considerations, the procedure indicates that court models may be adapted to decision making on technical matters in a workable, effective way.

THE SCIENCE COURT EXPERIMENT IN MINNESOTA

Ronnie Brooks of the Upper Midwest Council, Minneapolis, said that attempts to resolve conflicts through rational processes have been central to the evolution of government. Political systems have been designed and modified to enable society to achieve acceptable solutions to disputes caused by competition between alternative value systems. During the earlier phases of this evolutionary process scientific facts were rarely in dispute but now they have become staples of the political process. The judgment of the scientific community is no longer accepted regarding disputed facts or the conclusions to be drawn from them. Nor is the scientific community even unified any longer, for it has come to depend on the very institutions competing within the political system.

The problem of resolving controversies with scientific or technological components is compounded by the communication gap that exists between the scientific community and the public realm, which includes representative government. Most of the actors in the public sector are as unaccustomed to dealing with scientific and technological information as is the scientific community in analyzing the broad social and political implications of its work. What all of us must attempt to do is to explore alternative ways to bridge that gap—ways that are consistent with the methodological standards of the scientific community as well as with our democratic political traditions. The task is not an easy one. We have very little experience in this kind of problem solving and, it seems, very little time to learn. Disputes over scientific and technological issues which have major impact on our society are occurring at a rapidly increasing rate, and our inability to resolve them in an appropriate manner is very costly in both social and economic terms.

Since we have so little experience in this area, this chapter is an attempt to share the history of one example of using an innovative method of resolving a technological dispute arising in the public sector in Minnesota. Hopefully, sharing the Minnesota experience with a conscious attempt to experiment will suggest improvements which can be made in the process of resolving these kinds of disputes in the future.

Ronnie Brooks began by saying that the attempt to use a science court mechanism to resolve a dispute over the construction of a high-voltage transmission line in Minnesota was not successful. Understanding why it was not successful is essential to refining the science court concept and further defining its applicability to real conflicts. Moreover, evaluating the Minnesota experience points out the need to assemble a range of alternative methods of resolving disputes over scientific and technological issues arising in the public sector, methods suited to the nature of the particular facts in dispute and the character of the disputants.

Several caveats are necessary before we begin to discuss the science court experiment in Minnesota. First, the dispute over the construction of a high-voltage transmission line in west central Minnesota was more than a technological dispute. Like so many other disputes, the scientific and technological issues were inflated and distorted and confused with value considerations before they could be resolved. Second, the author was a participant in the efforts to resolve the dispute beginning in January 1977, representing a governor who had just taken office. However, the dispute began several years earlier and several of the major actors in the decision-making process had changed. Finally, the dispute is still somewhat with us so it is difficult to reach many firm conclusions or to tell whether or not any of the lessons learned have any general applicability. Moreover, in recalling a series of events in which one was a participant, it is difficult to remember which beliefs or perspectives were held at any given time and which have emerged as the result of "Monday morning quarterbacking."

The case history begins in 1972 when two rural electric power cooperatives began discussions of a jointly owned generation and transmission line project. The following year they commissioned first a feasibility study and then engineering work on the project which involved the construction of a coal-fired generating plant in North Dakota and power transmission by high-voltage line into Minnesota. During this period, the Minnesota State Legislature passed several pieces of environmental protection legislation. For the purposes of understanding this case, the most important of these new laws was the Power Plant Siting Act, enacted in 1973, which gave to the state, through the Environmental Quality Board, the authority to determine the route of all high-voltage transmission lines. The Board was composed of several state agency heads, a representative of the governor, and a minority of citizen members also appointed by the governor.

Before 1973, no one in government at any level had any substantial authority over where power companies built plants or routed lines. The companies, which had the power of eminent domain, decided where they wanted to go and dealt quietly with each affected landowner while the threat of condemnation hovered over the negotiations. The result was a series of routing decisions based exclusively on economic considerations, oblivious to environmental or social concerns. The new law changed that and outlined a siting process which provided for substantial public input and ultimate state control through the Environmental Quality Board. The process required that the Minnesota Energy Agency first grant a certificate of need to the project. Then, as the process existed in 1973, there were two six-month steps to the selection of a route. The first step was to select a corridor, or study area, up to 20 miles wide and within which a number of route alternatives were possible. The second six-month step culminated in the selection of a route within the designated corridor. The emphasis throughout the siting process was on public involvement and rational evaluations of the alternatives. This was accomplished through the creation of citizens' committees, numerous information meetings, public hearings, media publicity, and direct mail to affected landowners. In addition, an environmental impact statement was prepared on the route alternatives. The final decision by the Environmental Quality Board was based on the public hearing record, the evaluation of route alternatives, and on the citizens' committee recommendations.

Cooperative Power Association (CPA) and United Power Association (UPA), the two power cooperatives, had begun the project on their own prior to the enactment of the new law and had run into considerable local opposition. It was to avoid meeting this opposition that they decided in 1975 to waive their right to follow the law in effect at the time they initiated the project and to have the Environmental Quality Board site the line under the provisions of the 1973 law. The board accepted the application for a transmission line route and began the required process, selecting for study two corridors in addition to the two proposed by CPA-UPA. A 47-member citizens' committee was appointed in April 1975 and the hearings began. To give some concept of the scope of the process, note that nine public information meetings were held in June of that year and 11 formal public hearings were conducted in July and August. During the hearings, which produced some 1,800 pages of testimony and exhibits, a great deal of testimony was received on scientific and technical questions relating to health and safety, such as the effects of exposure to ozone concentrations and electric fields. The board also requested and received reports on undergrounding and on electrical environmental effects.

The independent hearing officer in the case recommended the corridor suggested by the staff of the Environmental Quality Board which was, in essence, half of the CPA-UPA preferred corridor and half of the citizens' committee preferred corridor. The officer also recognized the concerns expressed over health and safety and made several recommendations to ensure safe operation of the line, recommendations which were ultimately included in the construction permit issued to CPA-UPA. After the corridor was selected in October 1975, the study process began again in preparation for the selection of a route within the designated corridor. CPA-UPA applied for a route and the Environmental Quality Board identified 80 segments for study. Again, a Citizens' Route Evaluation Committee consisting of 49 people, most of whom lived in the study area, was appointed. Twelve more information meetings (one in each county the line would traverse) and 12 more formal public hearings were held.

The hearing officer's recommendations included construction permit conditions on safety issues as well as on the route, and on June 3, 1976, after hearing final arguments from various organizations, the board accepted the hearing officer's recommended route and issued a construction permit for the line.

The conditions of the permit were stringent and called for monitoring of the ozone generated by the line to ensure compliance with existing health and safety standards. The permit also mandated numerous special actions by the power companies to reduce electric field strength under the line. These steps did little to mollify many of the landowners who continued to oppose the line and who, for numerous reasons, harbored very hostile feelings toward not only the power companies but also the administrative agencies that had participated in siting the transmission line. Protests against the line continued and lawsuits were filed. The controversy remained prominent when, in December of 1976, a new governor took office in Minnesota. Shortly after assuming office, in an effort to soothe the volatile tempers that threatened to erupt into violence, the new governor made an unplanned and unannounced trip to the center of the area affected by the project and met with protesting farmers in their homes and local meeting halls.

Those meetings established an atmosphere which we hoped would increase the likelihood of reaching some resolution to the issues. The farmers were pleased that the governor took time to listen to their concerns and it appeared that the level of emotion had been lowered adequately to permit a return to rational discussions. What the trip had also done, however, was substantially raise the protesters' expectations of getting some satisfaction from the state's governmental agencies. As another result of the governor's trip, joint legislative hearings on the powerline issue were held in an attempt to clarify the issues in dispute. The hearings were useful in that effort and the primary concerns aired by the protesters at those hearings were health and safety, compensation to landowners, and preference given in the law to preserving wildlife lands over agricultural land in the siting process.

A month of attempts by the governor's office to bring the power cooperatives and the protesters together to negotiate the points of difference was unsuccessful. Each side seemed determined to manipulate the scheduled sessions to its own advantage, either by playing to the media or by stalling pending an anticipated victory in the courts. Finally, the governor proposed bringing in an outside mediator and, to avoid the threatened violence, asked the power cooperatives to postpone construction activities along the proposed route pending the outcome of mediation. Reluctantly, the cooperatives agreed. A mediator from the American Arbitration Association trained in the resolution of community disputes came to Minnesota in early February and held exploratory discussions with both sides. Discussions were generally unproductive in terms of making progress toward resolving the dispute, but they did reconfirm the conclusions reached after the legislative hearings.

Since the inability to resolve the powerline controversy was very costly in political as well as economic terms, the governor's office tried to select the most appropriate method for addressing each of the issues. The issue of compensation was addressed both by the cooperatives in their negotiations with landowners and also by a bill passed by the legislature which provided additional annual payments to farmers out of property-tax revenues as compensation for lines and towers on their land. In that same legislation, the siting process was modified to give greater weight to agricultural impacts. The questions relating to administrative procedures in the siting process were addressed in numerous lawsuits culminating in a unanimous decision by the Minnesota Supreme Court in September of 1977 in favor of the power cooperatives, affirming the actions of the Minnesota Energy Agency and the Environmental Quality Board.

The remaining issues of health and safety, since they were essentially scientific in nature, were deemed appropriate for submission to a science court. The concept of the science court had been put forward by the governor's office in February of 1977 as a way of submitting the technical issues still in dispute to a neutral body composed of technically competent professionals who would render judgment on the relevant questions of fact. Both sides in the dispute would be represented in the adversary process prior to decision; both would participate in the presentation of their arguments. From February of 1977 through March of 1978 we actively, though sporadically, pursued the establishment of a science court to deal with the issues of the health and safety effects of high-voltage transmission lines. The idea was extremely popular at first and received a lot of support from the general public and the media.

During that year in which we pursued the science court experiment, the political situation was volatile. The strength and nature of the activities of the opponents to the line varied, and the major problem we had was that of identifying any stable leadership in the protesting group. We would meet with one group to explain the nature of the science court concept and the following week others would emerge as the group's representatives. Meetings with large groups degenerated into media events unconducive to negotiation and no workable or consistent channels of communication were established. For a while, the major obstacle to establishing the science court was the protesters' demand for a construction moratorium until the panel rendered its decision. The power cooperatives were unwilling to endure additional costly delays of construction since they had already surmounted all the administrative and legal hurdles. At one point, in November of 1977, the governor met with six representatives of the protesters, got them to drop their demand for a moratorium on construction activities, and secured formal agreement to participate in the establishment of a science court along the lines outlined by Kantrowitz. But by the time the cooperatives had agreed, one month later, the leadership of the protesting group had been replaced and the agreement was also replaced with a list of new demands significantly expanding the scope of the court's inquiry and, once again, requiring a construction moratorium. As we had begun to exhaust the agenda by making progress on each item, new items were added until it became clear that the prime issue was no longer any of those that had been identified, but rather "no line at all under any conditions at all."

The protesters by this time had been joined by a loose coalition of antinuclear and anticorporate power groups and some students from the cities. The nature of the protest activities changed and escalated to the point where the governor authorized sending up to 175 state troopers to the affected area to protect the construction workers. Violence was a reality. Negotiations on the science court essentially fell by the wayside when the protesters proposed a radically different structure for the court, one which made the governor the only judge, instead of a panel of experts, and gave him power beyond the scope of Minnesota law, and widened the agenda far beyond issues of fact that the science court concept was designed to resolve. The qualities of neutrality and professionalism were gone.

The science court experiment originally seemed both appropriate and exciting. As it turned out, it fell victim to events and antagonistic attitudes. During the effort it was repeatedly suggested that scientific facts and technological judgments could not be separated from policy issues in the political process. The judgment of matters of fact was well underway before the science court idea was introduced. Nor was there advance agreement that its outcome would be accepted. The question of how best to apply the science court in the context of public disputes remains unanswered.

SCIENTIFIC DISPUTES AND SOCIAL DECISIONS

John Bailar III of the National Cancer Institute observed that major societal questions involving scientific matters tend to have several common characteristics. First, none of these matters can, or should, be treated as a matter of simply discovering the "truth." Such questions are likely to involve values as well as facts, and sometimes the scientific facts become almost irrelevant. This is especially true when the matter involves very large societal investments or risks, with some promise of major gains or losses for a lot of people. For example, opposing sides on environmental issues are likely to reflect value orientations much more than factual questions, as in how a virgin forest is to be regarded or how desirable it is to create jobs. Unfortunately the parties to such disputes sometimes prefer to be seen as fighting for scientific objectivity, so that all the discussions involved become distorted. Such tactics can have immensely harmful long-term effects. Each time the public learns that facts and issues are being distorted in the name of science, they will see additional reason for believing that science itself is faulty, and that scientists have no better claim to credibility than have lawyers, politicians, corporate executives, or consumer advocates. And the public will be right, to whatever extent we as scientists permit or contribute to this abuse. We need some mechanism to objectively separate questions of fact from questions of values.

A second common point is that as a dispute ripens it becomes harder to find disinterested experts, though there may be a growing group of persons who pretend to be both disinterested and expert. The pool of scientists with a deep understanding of any disputed point is likely to be small, and each of them may have only one really good chance to affect the course of affairs, when he or she first expresses a position publicly. Later effectiveness is likely to be impaired because of a loss of credibility among persons leaning toward one side or the other or because of a need to respond defensively to criticism of views already expressed. Thus we need some mechanism of conflict resolution that will make best use of all available scientific talent, including persons already fully committed to one or another view.

Third, either the controversy or its resolution may have widespread and unexpected effects on other matters. Part of this diffusion of effects will be a natural outcome of the initial controversy and part will be because one or both sides find advantages in linking the dispute to other matters. It would be helpful to have mechanisms that will objectively separate the major issues so that the specific scientific questions at issue can be examined in a complete and undistorted context.

Fourth, the biggest scientific problems or uncertainties may have to do with assessing and interpreting the possible range of probabilities that certain effects will follow certain decisions, rather than with just giving a best figure. In statistical terms, estimates of mean effect may be less important than estimates of variance. While there may be good agreement that some untoward event has a likelihood of 10^{-6}, appropriate societal actions may depend critically on whether uncertainty about that estimate can be expressed as a range of 10^{-5} to 10^{-7} or 10^{-3} to 10^{-9}. We will rarely have precise estimates of either risks or benefits of some great decision; parties to a dispute will pick one or another extreme. We need objective mechanisms to consider the spectrum of possible outcomes and to assess the likelihood that each possible estimate is correct.

Next, one side or the other (sometimes both) will complain bitterly that its views have been misrepresented to the public. While this is sometimes true, it is not true as often as many scientists believe, and when it is true it is our own fault in nearly every case. Many scientists are wary of the press and of the news media generally. That is a mistake; the media can be very helpful, as they employ many dedicated professionals who want to do the best possible job. But their help must be enlisted by patient, honest expositions from the technical community.

Scientific disputes can arouse very strong emotions and provoke strong opposition. And while the basis of partisanship in scientific matters may be less manifest than in disputes of other kinds, the exponents of organized interests will act as vested interests have always done. Right or wrong, such interest groups involve themselves with a lively spirit of self-justification, believing that ends justify means and that their opponents deserve whatever can be done to them. We need mechanisms for resolving disputes that will make these groups play a more creative, less harmful part.

Careful consideration of these points suggests that it might be genuinely valuable to institutionalize new means for assuring proper consideration of scientific matters bearing on the great pending decisions before society. In particular, we should consider the establishment of formal, permanent mechanisms to identify the real scientific issues, to evaluate the state of our knowledge, and to develop objective assessments of the degree of uncertainty regarding any estimates of risk or benefit. While such formal mechanisms might take a variety of forms, their proceedings should be open and structured to bring specific, scientific issues into focus. They should lead to formal statements of findings, but not recommendations based on them. The managers of these proceedings should be professionally trained for the task, free of other responsibilities, and disinterested in the outcome. Recommendations for administrative action or political positions should be drawn by agencies or individuals other than those participating as advocates during the proceedings.

Consensus Meetings at the National Institutes of Health

In the consensus development meeting procedure in use at the National Institutes of Health (NIH) adversarial relations are muted. As the word consensus implies, the purpose of the meetings is to seek the largest possible area of common agreement, and this is by no means always possible. Still, the process met with enough success to result in the establishment (within the office of the director of NIH) or an organizational unit equipped with written procedures for identifying questions suitable for the consensus development approach. Three dozen meetings have been held or scheduled within the short space of a year and a half. This very rapid development testifies to the need for a formal mechanism to examine unsettled scientific questions.

As it has evolved the consensus procedure has been employed not alone to aid decisions in the fact of controversy but also to serve as the context for technology assessments. More recently it has taken on the function of linking the development of new knowledge to subsequent steps needed for its dissemination, such as questions of how and when to put the findings of medical research into general application so that the transfer to medical practice is neither too soon nor too late and occurs smoothly at the chosen time.

The Value of Confrontation

The consensus meetings do not resemble cross-examinations in a court of law. Views are presented to a panel of experienced generalists from an appropriate spectrum of disciplines, employing an informal question procedure during presentations. One may wonder if the scientific community is not too insulated from more strenuous procedures of argument, such as cross-examination. It is too easy to say that truth will out (in science anyhow), or that civilized people just do not engage in public argument. Nor is it enough to believe that we are all after whatever the truth turns out to be, for the truth may out only after large monetary outlays or after a major disaster. Argument in public before a court of peers might be a much more orderly and civilized process than the method to which many now resort, or charges and countercharges in the newspapers. Scientists are just as likely as anyone else to be rogues, fools, Machiavellians, or self-seekers and controversies mixing science with values tend to bring out the worst in them. If any special virtues attach to scientists, they are collective rather than individual. Direct confrontations could assure fairness and expose obvious grandstanding.

The assessment of facts needs to be separated from the process of making decisions based on the information. Scientists are not trained in the application of value judgments and are rarely in positions conferring responsibility for outcomes. Government procedures whereby experts are asked to produce both findings and recommendations are obviously deficient. Despite the frequent written warnings against automatic linkage between assessments of matters of fact and their integration with values, our practice remains unchanged. One exception worthy of note is embodied in Public Law 95-623, which makes the director of NIH responsible for forwarding to the director of the National Center for Health Care Technology lists of medical advances growing out of its research programs and ready to be considered for wider application. These lists will be based on the findings of consensus development meetings.

The advantages of a consensual process over a science court are partly due to the slow, ponderous mechanisms the latter must employ. It is not necessary to structure an artificial process whereby positions are elaborately worked out for presentation. The Minnesota experiment indicates how shrewdly opponents can exploit the careful stages of an elaborate process. Finally, the consensus approach seems better designed to identify areas of genuine uncertainty, which is a helpful means of reducing controversy. We should distinguish fact finding from conflict resolution and decision making and tailor our procedures accordingly.

NOTES

[1] Arthur Kantrowitz, "Controlling Technology Democratically," *American Scientist* 63 (September-October 1975); Arthur Kantrowitz et al., "The Science Court Experiment: An Interim Report," *Science* 193 (August 20, 1976); and Arthur Kantrowitz, "The Science Court Experiment: Criticisms and Responses," *Bulletin of the Atomic Scientists* 33 (April 1977).

[2] A complete description of this evaluation process and its results can be found in United States Department of Energy, "An Evaluation of Alternate Magnetic Fusion Concepts 1977," DOE Report DOE/ET-0047, 1977.

18
Communications
Michael Tyler

One by one different applications of electrical energy have succeeded one another on the stage of attention since Edison: first lighting, then mechanical uses of electricity, and now applications where electrical energy serves only to carry information. The briefest listing of such applications as telephones, computers, and television suffices to show how profoundly information technology has modified life and work, especially in the years since 1945 as one industrial country after another has reached the point of near-universal availability of telephones, radio, television, and, most recently, business computing. The impact of new electronic technologies is likely to widen as nonelectric means of communication come to employ them, in check payments, postal service, and newspapers.

The growth of technological capability that has brought about these major changes in a few decades shows no signs of slackening. The need for a better understanding of the forces at work, the way they mold economic, social, and political processes, and resulting issues and policy choices has never been more apparent. Our choices are made more complicated by the pervasive and poorly understood consequences of policy choices made or evaded 20 or more years ago regarding earlier innovations such as television. Better knowledge is needed as the basis for present-day policy making and not nearly enough social, economic, and behavioral research has been undertaken.

THE COMMUNICATIONS REVOLUTION AND THE QUALITY OF LIFE

Even more striking than the extent of the research that has been done on the impacts of telecommunications is the realization that gradually dawns during exposure to this field that very little indeed is yet known about some of the most important problems. To choose just two examples: we do not know whether to expect and plan for very large changes in the pattern of urban settlement and land use resulting from new patterns of work or how to take communications capabilities into account in considering the future of the airline industry. Some key trends may be identified. The cost of information processing is falling rapidly and the processing of text, data, and graphics by logic circuits can be widely dispersed throughout society, even to the level of building clever controls into sewing machines or toys. The cost of electronic memory is also falling rapidly. Less dramatic but still important is the decline in the cost of telecommunications, that is of transporting information over long distances. As all this has been taking place the costs of nonelectric media have been stable or, more often, rising.

The problems facing the traditional information media are not matters of private costs only. Social costs in the form of resource depletion, environmental damage, and urban problems are a major problem. Every issue of the *New York Times* involves consumption of immense quantities of wood pulp. Moving passengers by air or by automobile is one of the most wasteful of all forms of energy use. In 1970 it accounted for 25 percent of energy consumption in the United States and is also responsible for noise, air pollution, and injuries on a huge scale.[1] By comparison electronic technologies are generally benign and concern about the effects of microwaves or of excessive exposure to cathode ray tubes is low-key. These technologies are on the whole harmless and resource conserving, make only minimal demands on energy and scarce materials, and have few adverse environmental impacts. This lack of conspicuous side effects seems to some of us in the field a paradoxical disadvantage. If every telecommunications installation had a conspicuously smoking chimney, the importance of the issues raised by the electronic information industries would be more apparent to politicians and policy makers. Even so, the intangible but immense impact of information technologies on everyday life and on national and international policies is beginning to be felt, as the recent publication of several very useful books serves to show.[2]

Communication and Information—New Applications of Technology

Trends in the basic technologies of a communication system, such as memory and processor power, cause their social consequences only as they are introduced generally. Consider the main avenues of innovation as the user might view them, in terms of service attributes. Telecommunication services may be used for one-to-one communication as opposed to one-to-many; one-way and two-way; video and audio; simultaneous and delayed action; fixed location as opposed to mobile; and many more. We might consider television as a one-to-many, one-way, audio-video, simultaneous, and mobile communication mode, with only a very limited degree of selectivity. Program originators do much more selecting of program content than do viewers. In many ways, the telephone service lies at exactly the opposite end of the continuum of service possibilities. For example, television has full-motion video capability, while telephone offers low-fidelity audio only; basic telephone service is a one-to-one switched service, whereas television is one-to-many and unswitched; the originator (largely) selects the content of television programming, while "content" in a telephone conversation depends on both parties; and telephone is two-way and broadcast television is one-way.

Why is there so little, apart from radio, in the spectrum between these two extremes? The answer seems to be largely technological and economic, although there are some possible services that may have inherently low utility in most applications. The videotelephone is almost certainly a case. At one extreme, the traditionally high fixed cost of television programming and distribution tended to "freeze" broadcasting into its familiar pattern, as did the economic and technological barriers to greater variety and to two-way operations. Now both technological change, reflected in falling costs and in technological innovations such as interactive cable television, and social innovation in the world of local and special-interest broadcasting are beginning to break down many of the traditional assumptions about broadcasting. The development of telephone technology, too, is converging on the same "middle ground." For many years the economic superiority of radio transmission for broadcasting purposes obscured the potential of switched systems such as the telephone network, although in the early days telephone broadcasting applications such as Budapest's "talking newspaper" of the 1920s were widespread. The appearance of versatile automatic telephone networks as a result of applying computer ("stored program control") technology and low-cost storage has greatly increased interest in unorthodox applications such as small-group broadcasting or recorded-voice "publishing."[3]

There are certain dimensions along which television and the telephone do not seem opposed. Neither medium in its conventional form is a significant channel for textual information. Both media are synchronous so that all messages are, from the recipient's standpoint, ephemeral and cannot be recalled and viewed again. Increasingly, of course, additional capabilities for recording can be provided by new terminal equipment. Within the telephone network recent innovations seem more significant. Cheaper and more versatile transmission technologies are being introduced, such as digital techniques, packet switching, and optical fibers. Processors capable of "intelligent" switching and other functions may now be controlled by the programs they transmit. Low-cost, high-capacity memory devices employing semiconductor and magnetic bubble technologies are extensively available. "Intelligent" terminals exploit the capabilities of integrated circuits and microprocessors.

Striking new developments in service are emerging, such as electronic mail and funds transfer. Electronic handling of text and graphics now may be found in many offices, providing a comprehensive system of aids to the information worker and enlarging opportunities to work at home. Teleconferencing has become an alternative to the wasteful practices of routine business travel. Low-cost electronic publishing systems such as the United Kingdom's "Prestel," capable of using television screens to present user-selected text and graphic information, are developing as the basis for a worldwide electronic information marketplace.[4]

Two questions dominated our discussion at the Edison Centennial Symposium. Will these new communication media really be adopted and widely used and when? When their impact is felt, what will the consequences be for society? The implementation of even a small part of the potential for communication facilities offered by the new technologies would fundamentally transform society's infrastructure for communications. Whenever this has happened in the past, as in the introduction of canals in eighteenth-century England or the advent of the railway and steamship in the nineteenth century, immense social, economic, and environmental changes resulted.

Work and the Workplace

It is easy to forget how fundamentally economic life has been transformed in a few generations, especially since 1950. Most people now work neither in the extractive sector of agriculture and mining nor in the manufacturing sector but in information-processing activities. This change reflects both the massive progress of productivity in the sectors serving our material needs and the increasing centrality of knowledge and information to the entire productive process.[5]

Because of the trend to information processing in the workplace we considered the changes likely in the office environment rather than those that will undoubtedly also occur in manufacturing and extractive industries. Perhaps the most salient fact about the office environment is how little it has changed. The technologies that presently dominate the office are so familiar that we hardly think of them as technologies. Few of them date from the last 40 years, the most notable exception being the electrostatic copier pioneered by Xerox. Technologies like paper files, memoranda, letters, typewriters, and card indexes have shown little or no change. One notable consequence is the very low level of capital per worker, a few hundred dollars or, if buildings figure in the calculation, a few thousand. Will this difference from the capital invested per worker in manufacturing remain unchanged? Almost certainly not, because office productivity is becoming a central concern of management and the means to achieve improvements in staff productivity are becoming generally available.

Word-processing typewriters seem to be appearing in offices in a piecemeal fashion but more comprehensive adoptions of office technology are imminent, taking place by single major changeover akin to the conversion of manual accounting to computer operation. New technologies will surely make working in an office a very different experience in the future. Will increased productivity free staff time for more pleasant, creative tasks? Will clerical workers experience losses in autonomy as they find they are paced by machines rather than by co-workers? Will there be major problems of unemployment? What kind of psychological problems may arise from the more rapid flow of information from display screens to people? Such research as has been done with the few experimental advanced office systems now in use suggests that the answers to these questions depend very much on how the electronic systems are designed and on the kinds of organizational structures in which they are placed. Here we have an opportunity to influence the future for the better rather than passively forecast or blindly await its arrival. In order to do so, we should embark on social experiments with new technologies.

In assessing the potentials of new technologies and considering prospective problems we should be sensitive to their impact on the organization of society. Consider "remote work." Electronic office systems would free people from the need to carry out their information-processing work close to one another. Documents, data, voices, and images could travel many miles as easily as they now travel a few yards inside the office. Thus much work could be done from home or, if that proves unsatisfactory to some people, from some form of neighborhood work center. If this trend sets in widely differing patterns of human settlements might result.

Does the remote work idea have enough plausibility to merit future investigation? Although little evidence from research is available, our discussions suggested that a growing number of workers, primarily technicians and professionals, are already using their homes, even with the limited communication facilities now available.[6] A technology assessment carried out by the Stanford Research Institute in 1976 identified major economic benefits such as reduced transportation costs, major energy savings, and reduced office costs.[7] Obviously the present pattern of utilization of buildings in cities is extremely wasteful, as is our use of transportation systems at capacity for only three hours of the day. Socially it seems obvious a priori that greater locational freedom and choice of home and work patterns must be a benefit, but this may only be true in the absence of externalities in which some individuals' choice of work style harms others. Dispersal induced by telecommunications could accelerate the decline of city centers, with many attendant economic and social problems. Another problem would be increased development pressure on rural lands. Can we evolve far-sighted urban and regional planning policies to make the most of the opportunities these new technologies afford?

Communications in the Home and to the Home

The electronic communications revolution has so far affected the home in two respects, the television set and the telephone. We are still far from fully understanding and appreciating their consequences, as demonstrated by the continuing debate about the influence on children of televised violence. The social impact of the telephone has been much less extensively studied. We have only the vaguest idea of what part it may have played in the massive changes of the last fifty years. These two modes will soon be augmented by other means of bringing text and graphic communications into the home, substituting perhaps for existing media such as newspapers and magazines. Services for the sending and receiving of messages and financial transactions such as banking and purchases are expanding.

Electronic publishing for the specialist user is not new. Systems such as MEDLINE or the Information Bank of the *New York Times* are well established. What remains in question is how the next step will be taken, to a cheap, easily understood "Model T" version of electronic publishing suitable for a wider market among unsophisticated users, principally at home. The implications of innovative uses of a medium that can bring a potentially very large body of information into a home under push-button control are presently difficult to foresee. We tend to assess them in terms of categories derived from such existing service patterns as the "Yellow Pages." Undoubtedly many new formats for education and entertainment will emerge. Stimulating certain kinds of innovation may be crucial to the eventual success of the new media, especially the realization of their employment-creating potential.

Interactive message services will bring into reality such concepts as shopping from home. A range of more socially oriented services could also be provided to the home or through community facilities such as clinics or libraries, using various combinations of new and old telecommunications media. Areas of special interest include schemes to improve access to expertise through remote professional consultations such as "telemedicine" or attempts to improve citizen participation through enlarging opportunities to receive and discuss information.[8] This idea brings us back to the evolutionary nature of telecommunications impacts by encouraging us to recognize how much of the technical infrastructure needed for a revolution in communications is already in place. Many cable television systems, for example, already carry some community-oriented local programming and can be adapted at an acceptable cost to carry two-way interactive services. In Reading, Pennsylvania, for example, a project initiated by New York University's alternative media center used new technology to stimulate community activities and involvement on the part of older people, with such a high degree of success that the system is now fully financially supported by the local community after its federal funding came to an end, a rare occurrence for pilot projects. It is clear, then, that the proliferation of new electronics-based communication media in almost every area of everyday life is both possible and, in many areas, likely. The major need is to develop policies to maximize the benefits of the new media, skirt the potential dangers, and make the most of the opportunities they offer.

NETWORKS AND COMMUNICATIONS ECOLOGY

Ithiel de Sola Pool, Professor of Political Science, Massachusetts Institute of Technology, recounted aspects of the history of electricity serving to establish the metaphor by which it has come to be seen as the nervous system of society. This image had two distinctly different meanings. By the first, all society would be integrated into a unified whole by the network of wires. The rival conclusion was that electricity would allow a dispersal and decentralization of society, away from the densely populated cities and crowded factories of the age of steam.

Nathaniel Hawthorne's character Clifford expressed the first view in *The House of the Seven Gables*. "Then there is electricity, the demon, the angel, the mighty physical power, the all-pervading intelligence. . . . Is it a fact that by means of electricity the world of matter has become a great nerve, vibrating thousands of miles in a breathless point of time? Rather, the round globe is a vast head, a brain, instinct with intelligence!" The extreme view of the electric network as unifying all society, in fact a caricature of it, was Lenin's, who saw electricity as turning the whole Soviet system into one gigantic factory. He used the words "large-scale machine industry" and "electrification" more or less synonymously.

The reverse view of electricity was characterized as more accurate, as a force against gigantism. It was expressed before the Institution of Electrical Engineers in London in 1889 by the then prime minister, the Marquis of Salisbury.

> The electric telegraph has achieved this great and paradoxical result: it has, as it were, assembled all mankind upon one great plane, where they can see everything that is done, and hear everything that is said, and judge of every policy that is pursued at the very moment these events take place. And you have by the action of the electric telegraph combined together almost at one moment, and acting at one moment upon the agencies which govern mankind, the opinions of the whole of the intelligent world with respect to everything that is passing at that time upon the face of the globe.

The same view of electricity as an agent of decentralization and diffusion is found in H.G. Wells' *Anticipations*, the book he wrote at the start of the twentieth century outlining what he envisioned over its course. He described the "deliquescence of the social organizations of the past and the synthesis of ampler and still ampler and more complicated social unities." Wells understood the complex dual nature of the growth of ever larger systems: on the one hand larger and more complicated units and on the other, the growth of dispersed units among them. In a chapter entitled "The Probable Diffusion of Great Cities" he described the phenomenon that Jean Gottman has labeled Megalopolis. Electricity plays a large causal role in this with the streetcar and telecommunications allowing dispersion. Also he describes a new lifestyle in which electric appliances allow for the disappearance of servants. Like Edward Bellamy before him, he described convenient homes with numerous electrical and mechanical appliances. He also described the growth of the information-processing class, what Marc Porat styles the growth of the information society. This view of the impact of electricity was shared by Lewis Mumford in *Technics and Civilization* in 1934.

In retrospect it seems clear that the views of Salisbury, Wells, and Mumford were far closer to the reality of the effects of electricity than those of Lenin, though both stressed a part of the complex reality. The growth of the system led to increased diffusion of centers of activity rather than to increased centralization, though within a vast connected system. Of course, human choice has a role as well. The Soviets, given their view of electricity, acted fastest in the development of a unified grid. Less centralistic ideologies relied upon electricity in seeking to realize a vision of smaller, more independent units. Such a view was expressed by Frederick Perrine in an essay on electrical engineering and social reform published in *Electrical Engineering* in 1894. He foresaw extensive suburbs, made possible by the streetcar, an end to the pollution of steam locomotion, and a return to work at home. The spirit of electrical engineers, he believed, was "pressing toward social reformation." Perrine recalled that the first electrical congress in Paris had included, sitting among the delegates, many whose early manhood had been spent in the St. Simonean community preaching the doctrine of "the world's reform." That early, intense generation was intent on centralization, prompting the criticisms of Friedrich von Hayek in *The Counter Revolution of Science*. After Perrine, engineers looked foward to utopias that can still inspire us, of a society that diffuses great cities into the countryside, allows work to be done in units of human size, and allows us all to communicate with one another and not just be the objects of communications addressed to us by others.

NOTES

[1] Some comparison between transportation and telecommunications can be found in Stanford Research Institute, "Technology Assessment of Telecommunications /Transportation Interactions," 3 vols., 1977. Also M. Tyler, M.C.T. Elton and J.A. Cooke, "The Contribution of Telecommunications to the Conservation of Energy Resources. A Background Paper for the Organization for Economic Cooperation and Development," OT Special Publication 77-17 (U.S. Department of Commerce, July 1977).

[2] A particularly useful overview book providing a map of the telecommunications policy field is Glen O. Robinson, ed., *Communication for Tomorrow* (New York: Praeger, 1978).

[3] An excellent survey of trends and issues from a predominantly historical perspective is provided by Ithiel de Sola Pool, ed., *The Social Impact of the Telephone* (Cambridge, Mass.: Massachusetts Institute of Technology Press, 1977).

[4] See, for example, M. Tyler, "Videotex, Prestel, and Teletext: The Economics and Policies of Some Electronic Publishing Media," *Telecommunications Policies* (March 1979).

[5] M. Porat, "The Information Economy," Stanford University Institute for Communication Research Report no. 27, August 1976.

[6] R. Pye, "The Effect of Telecommunications on the Location of Office Employment," *Omega: The International Journal of Management Science* 4 (1976): 289-300; and J.B. Goddard and R. Pye, "Telecommunications and Office Location," *Regional Studies* 11 (1977): 19-30.

[7] Stanford Research Institute, "Technology Assessment."

[8] John Short, "Residential Telecommunications Applications: A General Review," Post Office Telecommunications Headquarters Long Range Intelligence Bulletin no. 6, London, 1977.

Index

Adams, Henry, 46
Bailar, John, 108-10
Basalla, George, 39-51
"Big science," 8
Bridge financing, 31
Brooks, Ronnie, 202-9
Brun, Thierry, A., 195-6

"Centennial of Light," vii, ix
Cities and Third World
    economic growth, 27
Civilization
    and ideology of energy use, 40
    and misconceptions of technology, 88
    equated with levels of energy use, 39-51
    moral dimensions of, 102-3
    problem-solving mission of, 124
Communications, workshop on, 212-19
Compassion, 57
Consensus meetings, 210
Creativity
    and style of institutions, 8
    as culturally determined, 13
    fluctuations in, 4-10
    hard work of, 57
    industrial innovation compared to, 5, 11
    in order to adapt technology to
        unconventional contemporary
        realities, 182-3
    need for corporate climate of
        confidence, 69
    organization for, in Third World, 33-4
Currie, Laughlin, 26

Decker, James F., 199-201
Democratic control of technology,
    workshop on, 199-211

Djojohadikusumo, Sumitro, 26-37
Dumont, René, 185-198

Eberhard, John P., 177-184
Economic system, worldwide revisions in,
    28-34
Edison, Thomas A.
    and Muscle Shoals project, 47
    "battle of the systems," 145-6
    concept of technology as a social
        system, 81-2
    general philosophical concepts, 87
    mathematical knowledge, 79
    technological capability, 144
    x-ray demonstration, 172
Eisenbud, Merril, 169-175
Electric power industry
    and technological needs, history of,
        141-56
    attempts in regional adaptations in
        Germany, 153-4
    difficulties in development in London,
        147
    prospective public informing role,
        94-108
Energy, workshop on, 129-140
Energy consumption
    equated with civilization, 39-51
    firewood, 36
    from planets in the Diesel engine, 195
    global level and economic well-being, x,
        28
    importance of coal as the swing fuel, 36
    role of technology, 73
    schools of thought in forecasting, 131
Environmental impact of technology, 36,
    198

Falcon, Walter, 194
Food and agriculture, workshop on, 185-198
Food production
  and social well-being, 111
  and Third World agricultural income, 26
  role of technology in, 74
  social assessments of, 185-198
  United States forecast as too low in 1922, xi
Ford, Henry, 47
Free enterprise sector
  difficulties experienced in applying technology to social needs, 71
  limits on decision-making power, 100
  morality of, 94-108
  need for cooperation with government, 68
Fritsch, Bruno, 136-7

George, Susan, 196-8
Government
  limits on decision-making powers, 100
  role in fostering technology, 74
  tensions with technology, 67-8
Growth, economic
  and American industrial innovation, 14-24
  dependence on energy, 137
  failure of distinction from environmental protection, 36
  importance for world aggregate demand, 28
  lack of limits upon, ix-xiv, 160, 188-9
  limits upon, 85
  need for interventions in, 163
"Growth of limits," ix-xiv

Häfele, Wolf, 129-40
Hambraeus, Gunnar, 3-13
Hannah, Leslie, 151-3
Handler, Philip, 109-125
Hansen, David H., 192-3
Hare, F. Kenneth, 157-160, 167-8
Health
  and social well-being, 110-11
  and world population, 165
  as defined by WHO, 171

Hoffer, Eric, 53-8
Hughes, Thomas, 141-151
Human factor and social predictability, 53
Human population and ecology, workshop on, 157-168
Husserl, Edmund, 87-8
Huxley, Aldous, 40

Industrial research
  impact of government regulation on, 20, 23
  institutional aspects of, 66-76
  long term versus short term, 18
  marginal rate of return on, 15-17
  United States government review of, 22-3
Innovation, technological
  and institutional sectors, 66-76
  economic determinants of, 17
  in rural development, 28
  in social systems context, 141-156, 212
  need for formal studies of, 75
  role of organizations in, 13
  urban aspects of, 178
Institutional change
  and negative feedback, 164
  in Third World economic structure, 26
  opposition of economists from industrial nations to, 33
  needs for in adapting technology to urban needs, 181, 219
  role in innovation, 74
International Institute for Applied Systems Analysis (IIASA), 131-6
Irrationality of the human condition, 55

Jevons, W. Stanley, 43

Kantrowitz, Arthur, 199-211
Kerala, population programs in, 167
Keyfitz, Nathan, 27

Levi, H.W., 137-9
"London problem" in electric power, 147
Lovins, Amory, 143

MacIntyre, Alasdair, 94-108
Malik, F. Bary, 34

Manpower in science and technology, 34, 69
Mansfield, Edwin, 14-24
Medicine and public health, workshop on, 169-175
Morrison, Philip, 61-5
Mrak, Emil M., 193-4
Multinational corporations. *See* Transnational corporations
Mumford, Lewis, 177, 198

Natural resource inventories, 36
Nuclear power, 91
 as a prototype of modern problems, 137
 controversies over, 118

Oil recovery, 69
Ostwald, William, 45

Packer, Leo S., 180-1
Perrine, Frederick, A.C., 219
Pollution of "affluence" and "poverty," 36
Pool, Ithiel de Sola, 218-19
Poor, absolute, in global economy, 26-8, 185-198
Population, human
 and resulting problems, 137
 ecological perspectives on, 157-168
 not a cause of the hunger problem, 197
Power plant siting, 202-08
Problem solving
 and absolute poverty, 26-8
 and positive conceptions of morality, 94-108
 and Renaissance science, 7
 and technological productivity, 66-76
 and urban context, 178-9
 basis of social needs for science, 34
 in energy, 129
 individual versus institutional aspects of, 109-110
 need to be cognizant of social aspects of, 152
  primary concern of civilization, 124
  technological systems solutions, role in, 141
 tendency to leave to the last minute, 162
Public attitudes
 as exemplified by the Edison myth, 80-1
 influence of a science court upon, 199-211
 need for process of choice, 94-108
 toward economics, 75
 toward science, 10, 193
 toward technology, 66, 84-5
Pure Food and Drug Act, 173

Ramo, Simon, 66-76
Ridker, Ronald G., 160-3
Rosenfield, Patricia L., 165-7

Salomon, Jean-Jacques, 77-92
Schnee, Jerome, 172-5
Science
 and social achievement, 109-125
 light of, 61-5
 needs for objectivity, 120
 plateaus in and slowing of technological innovation, 174-5
 small role in the ecological crisis, 168
"Science court," 199-211
Shamos, Morris, 171-2
Small business, role in innovation, 180
Smith, Robert I., vii-viii
Soddy, Frederick A., 46-7
Spitz, Pierre, 198
Starr, Chauncey, vii, ix-xiv
Steam engine, 41
Substitute division of labor, 32

Technological history and technical problems, workshop on, 141-156
Technology
 and employment, 13
 and environmental problems, 71
 and growth of industrial productivity, 14-24
 and human progress, 110
 and range of social options, xiii
 and status of women, 112
 and urban transportation, 70
 appropriate, 31-2, 34
 environmental impact, x, xii, 12

extension linkages to users of, 36
institutional factors in, 66-76
making man superfluous, 53
mineral resource recovery, xi
protective, 35
public health, contributions to, 169
public participation in, 90-1
reflexive, 156
social and economic myths of, 17-92
tensions with government, 67-8
underdevelopment, role in, x, 26-8
varieties of definitions of, 182
Tennessee Valley Authority, 48
Timbs, John, 42
Transnational corporations
and the world economic order, 30
conducting research overseas, 174
role in innovation, 22, 36

Tyler, Michael, 212-19

Universities, role in technology, 75
Urban development, workshop on, 177-184
Urban transportation, 70
Utopias, 219

Vallentyne, John R., 163-5

Water resources, importance for development, 27, 28
Weingart, Peter, 153
Whitehead, Alfred N., 61, 65
Wilheim, Jorge, 181-3
Women, role of, 112, 196-7

Youmans, Edward L., 45

About the Contributors

Chauncey Starr, Vice-Chairman of the Electric Power Research Institute, served as its founding president from 1973 to 1978. Following an industrial career in nuclear science and engineering he was Dean of the School of Engineering and Applied Science at the University of California at Los Angeles from 1966 to 1973.

Philip C. Ritterbush is an historian and interpreter of scientific ideas and their cultural affinities who has served as editor of the *Prometheus* series of policy studies of contemporary institutions and as planner of the Smithsonian Institution's international symposium series. He is director of a center for higher education programs which is developing new approaches to interpreting science and technology in general undergraduate education.

George Basalla is Associate Professor of History of Science and Technology at the University of Delaware. In books such as *The Rise of Science: Internal or External Factors?* (1968) and articles such as "Museums and Technological Utopianism" (in *Technological Innovation*, I. Quimby, editor, 1973) he has sought to maintain the fruitfulness of the social history of science as a source of critical ideas and policy insights rather than a closed scholarly field.

Sumitro Djojohadikusumo is Professor of Economics in the University of Indonesia. From 1973 to 1978 he served as Minister of State for Research in the Indonesian government. He is the author of *Science, Resources and Development* (1977) and several other books, including *Economic Problems of Indonesia* (1951). He spent the period from 1957 until his return to Indonesia in 1967 as an exile from his native country, working as an economic consultant.

René Dumont is an agronomist and exponent of socialist political thought who has written more than 30 books on the problems of agricultural and political development in tropical regions. From 1933 to 1974 he taught at the National Institute of Agriculture in Paris, where he held the chair of comparative agriculture and agricultural planning. His best-known books in English are *False Start in Africa* (1966) and *The Hungry Future* (1969).

John P. Eberhard is Chairman of the Board of Architectural Research Associates and formerly served as President of the AIA Research Corporation and as dean of the School of Architectural and Environmental Design, State University of New York at Buffalo.

Merril Eisenbud is Professor of Environmental Medicine and Director of the Laboratory for Environmental Studies in the New York University Medical Center. He has served for two years as Environmental Protection Administrator for the City of New York and for twelve years as Director of the Health and Safety Laboratory of the Atomic Energy Commission. He is the author of a monograph, *Technology, the Environment, and Human Health* (1978).

Wolf Häfele is Deputy Director of the International Institute for Applied Systems Analysis in Laxenburg, Austria. He is a physicist specializing in nuclear reactor applications. He is the author of a forthcoming book based on the International Institute of Applied Systems Analysis (IIASA) energy program which he directs—*Energy in a Finite World: A Global Energy Analysis*.

Gunnar Hambraeus is Professor and Managing Director of the Royal Swedish Academy of Engineering Sciences, Stockholm. He has served as Editor of the technical journal *Teknisk Tidskrift*, as a government adviser on research, energy, and environment, and as Secretary of the Swedish Technical Research Council.

Philip Handler, President of the National Academy of Sciences since 1969, is a biochemist and member of the faculty of Duke University. He is the author of numerous scientific papers and co-author of a biochemistry textbook. He planned and edited *Biology and the Future of Man* (1971), the National Academy of Science's report on the progress and applications of biological knowledge.

F. Kenneth Hare is University Professor and Director of the Institute for Environmental Studies at the University of Toronto. He is the author of *Physical Geography for Canada* (1953), *The Restless Atmosphere* (1953), and contributions to numerous books on climatology, geography, and education.

Eric Hoffer is an author whose first book, *The True Believer* (1957), was written during the depression years when he was a migratory field laborer. Following 1943 he worked as a longshoreman while writing such books as *Working and Thinking on the Waterfront* (1969). His most recent book is *In Our Time* (1976).

Thomas P. Hughes is Professor of the History of Technology at the University of Pennsylvania and Chairman of the Department of History and Sociology of Science. He is the author of *Elmer Ambrose Sperry: Inventor and Engineer*

(1971) and editor of *Changing Attitudes toward American Technology* (1975). A forthcoming book will treat the history of electric light and power systems.

Arthur Kantrowitz, Senior Vice-President of the Avco Corporation, is a physicist specializing in gas dynamics and ballistic and aeronautical applications. He serves as the Chairman of the Avco Everett Research Laboratory, Inc. His proposal for a "science court" has prompted widespread discussion of procedures for decision making and the resolution of controversies in which science and technology are involved.

Alasdair MacIntyre is University Professor and Chairman of the Department of Philosophy at Boston University. He is the author of numerous books on social philosophy, including *Marxism and Christianity* (1953), *Marcuse: An Exposition and a Polemic* (1970), and *Against the Self-Images of the Age: Essays in Ideology and Philosophy* (1971).

Edwin Mansfield is Professor of Economics at the University of Pennsylvania. His books include *The Production and Application of New Industrial Technology* (1971) and *Industrial Research and Technological Innovation* (1968). He serves many national and international technical agencies as an adviser on research support policy.

Philip Morrison, Institute Professor and Professor of Physics at the Massachusetts Institute of Technology, is an astrophysicist, exponent of arms control, and interpreter of the sciences for the wider public. He is the author of numerous scientific papers, co-author of a physics text, *Introductory Nuclear Theory*, serves as book reviewer for *Scientific American*, and has published both books and articles on arms control and other aspects of public policy, of which the most recent is *The Price of Defense: A New Strategy for Military Spending* (1979).

Simon Ramo pursued a scientific career in the physics of microwave radiation and was one of the founders of TRW, Inc., serving as its Executive Vice-President, Vice-Chairman of the Board, and Chairman of the Executive Committee. He has served as Chairman of the President's Committee on Science and Technology and in numerous advisory capacities in government and education.

Jean-Jacques Salomon heads the Science Policy Division in the Organization for Economic Cooperation and Development and is Professor of Technology and Society at the Conservatoire National des Arts et Métiers in Paris. He is the author of *Science and Politics* (1973) and serves on many international bodies devoted to the study of science and technology policies.

Robert I. Smith is Chairman of the Board and Chief Executive Officer of Public Service Electric and Gas Company, Newark, New Jersey. An engineer by profession, he served as Chairman of the "International Centennial of Light."

Michael Tyler is Managing Director of Communication Studies and Planning, Ltd. of London. He is an adviser and consultant to the British Broadcasting Corporation, the United Nations, and numerous foundations and expert bodies. His technical report, "Prospects for Teleconferencing," published by the British Post Office, has been widely studied.